Lipid Biochemistry:
an introduction

Lipid Biochemistry:

an introduction

M. I. GURR
A. T. JAMES

CORNELL UNIVERSITY PRESS
ITHACA, NEW YORK

CHEMISTRY

International Standard Book Number 0–8014–0652–8

Library of Congress Catalog Card Number 70–159485

Printed in Great Britain

Contents

Preface

The aim of this little book is twofold; first to aid students in learning about lipids, and staff in teaching a subject that they personally might perhaps find dull; and second to influence students towards research in this area. We do not intend to provide a research monograph. Frankly, we find the subject fascinating and hope to attract more workers into the great range of research topics that are lumped together under the title 'lipids'.

We are thus grinding an axe and have attempted to make the book a little lighter in tone than the average run of textbooks. Our own interests can be detected by the difference in detail of treatment — for this we make no apology, since we believe that a section written out of enthusiasm is better than one done out of duty. If we fail in these objectives, we shall have dissuaded biochemists, physical chemists and others from entering the field — the only benefit from this would be to lessen the pressure on the present workers from apprehensive scanning of the literature to see if they have been beaten to it! It's an ill wind....

December 1970   M.I.G.

Sharnbrook, Bedford   A.T.J.

Acknowledgements

It is a pleasure to record our thanks to the many people who have helped in the preparation of this book. Our colleagues, Drs E.F. Annison, B.W. Nichols, L.J. Morris, C.H.S. Hitchcock, R.B. Leslie, D. Husbands, R. Bickerstaffe, D. Howling, G.H. Hübscher J.N. Hawthorne and G.M. Gray and Messrs P. Harris, D. Brett, S. Hall and M.P. Robinson for reading and commenting on different parts of the manuscript. Their discussion has been invaluable; Miss C. Inchley for many arduous hours of typing and for deciphering our abysmal handwriting; Mrs. E.A. Gurr for designing and drawing some of the figures; Messrs H. Deakin, D. Chance and Miss L. Needham for preparing many photographs and diagrams. Permission to reproduce certain photographs and diagrams has kindly been given by Dr B.W. Nichols (Figs. 1.2, 1.3); Mr P. Harris (Fig. 1.9); Dr. F. Vandenheuvel and the American Oil Chemists Society (Figs. 2.1, 2.2); Professor F. Lynen and the *Biochemical Journal* (Fig. 2.5); Dr W. Stoffel and *Hoppe Seyler's Zeitschrift für Physiologische Chemie* (Fig. 2.16); The Elsevier Publishing Co. (Fig. 2.22); Dr. L.J. Morris and John Wiley & Sons Ltd (Figs. 2.23, 3.9); Dr. L.J. Morris and The Elsevier Publishing Co. (Fig. 2.24); Mr. R.J. Taylor (Fig. 3.4); Mr R. Kenworthy (Fig. 3.11); Professor J. Senior and *The Journal of Lipid Research* (Figs 3.12 and 3.14); Mr C. Smith and Mr R. Kenworthy (Fig. 3.13); Dr. V. Luzzati and the *Journal of Cell Biology* (Fig. 6.1); Drs A.A. Benson, R.M.C. Dawson, D.E. Green, The American Oil Chemists Society, The National Academy of Sciences of the U.S.A., and Academic Press Inc. (Fig. 6.3); Mr M. Stubbs (Fig. 6.4A); Dr D.E. Green and the *Journal of Cell Biology* (Fig. 6.4B); and Dr. A.F. Henson (Fig. 6.4C); Professor O. Westphal and *Angewandte Chemie* (Fig. 6.5a); Dr L. Rothfield and the *Journal of Biological Chemistry* (Fig. 6.5b); Dr J.L. Strominger and The National Academy of Sciences of the U.S.A. (Fig. 6.6).

1 Lipids: what they are and how the biochemist deals with them

GENERAL INTRODUCTION

The word 'lipid' (in several different spellings) has long been used to denote a chemically heterogeneous group of substances, having in common the property of insolubility in water, but solubility in non-polar solvents such as chloroform, hydrocarbons or alcohols. Adequate coverage of the whole spectrum of such fat-soluble substances is beyond both the scope of so short a treatise and the capabilities of the authors. We shall therefore narrow our definition to include only those compounds which are esters of long chain fatty acids. Therefore large groups of biochemically interesting lipids such as the steroids and terpenes will not be covered, although our definition necessitates inclusion of, for example, the sterol esters.

Because, in our definition, the unifying feature is the long chain fatty acid, we will start by elucidating the various classes of naturally occuring long chain fatty acids (chapter 2). The organic and physical chemistry of these compounds has been described in detail in a companion volume — F.D. Gunstone's 'An Introduction to the Chemistry and Biochemistry of Fatty Acids and their Glycerides' and chemical and structural aspects will be dealt with only in so far as it is necessary for an understanding of their biochemistry — anabolism and catabolism — which makes up the major part of chapter 2. The tremendous upsurge of interest in lipids over the past few years has been due mainly to the development of powerful new techniques for the separation, analysis and identification of these compounds whose immiscibility with water and rather similar physical properties had hitherto led to rather slow progress. We shall therefore end each chapter which concentrates on a distinct class of lipids with a section devoted to the most useful analytical methods, but devote the greater part of this present chapter to

1

general chromatographic techniques.

From there, it is a natural step to describe the different classes of lipids derived from fatty acids: the so-called neutral lipids (non-polar esters of fatty acids with the alcohols, glycerol, cholesterol, vitamin A and other long chain alcohols, chapter 3); the phospholipids (mixed-acid esters of fatty acids and phosphoric acid with glycerol or sphingosine, chapter 4) and the glycolipids (a heterogeneous class in which each member contains a sugar moiety, chapter 5).

Lipids rarely exist in an organism in the 'free' state but are more usually combined with proteins or carbohydrates as lipoproteins or lipopolysaccharides. These form the subject matter for chapter 6. Much could be (and has been) written about the metabolism of lipids in disease. Such topical subjects as coronary heart disease necessarily involve a discussion of lipid biochemistry. We cannot hope to cover much ground here but have thought it necessary to include sections dealing with various disorders involving malfunction of lipid metabolism in the appropriate chapters.

NOMENCLATURE AND STEREOCHEMISTRY

In any area of scientific investigation, the naming of compounds tends to develop rather haphazardly, each new substance being named according to the whim of its discoverer. Hence in fatty acid chemistry, many compounds have been named from the source of their extraction, for example, palmitic acid from palm oil. This results in a highly confusing and arbitrary system of jargon which puts great strain on the memory. Later there follows a period of rationalization when a logical system of chemical naming is sustained for the old trivial names. These chemical names often have the disadvantage of being rather long and cumbersome, and furthermore, if the naming committee is not truly international, more than one 'rational' system may develop side by side. As the authors have been involved in lipid research for a number of years, the use of trivial names has become second nature, and as they are much less cumbersome to write, we have tended to use them in this book. Therefore, when a certain compound is first introduced, its chemical name will be stated in parenthesis after the trivial name; subsequently, the trivial name only will be used. Furthermore, the very convenient shorthand nomenclature for fatty acids, which has become very popular in the field of gas chromatography, will be used wherever possible (see Chapter 2, page 21).

When a field has developed rapidly, substances which were originally thought to be pure, are later found to be mixtures of two or more rather similar components. Therefore a name which was once intended to denote a single compound eventually has to include a whole class of related substances. This difficulty arises in the case of the phospholipids where the term *Cephalin* includes both phosphatidyl ethanolamine, phosphatidyl serine and some other lipids. Confusion has arisen because some authors have used the term to mean simply phosphatidyl ethanolamine.

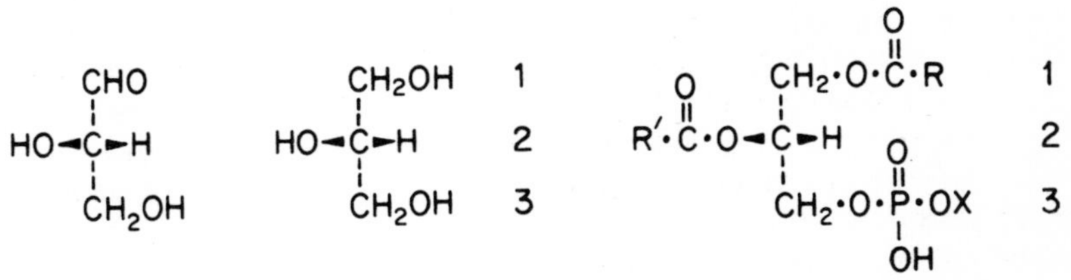

L-glyceraldehyde glycerol

L-3-phosphatidyl choline
D-1-phosphatidyl choline
1-2-diacyl-*sn*-glycero-3 phosphoryl choline

(X = Choline)

Fig. 1.1. The stereochemistry of phosphoglycerides

The terms *Cephalin* and *Lecithin* (phosphatidyl choline) will be avoided in this book (except at the first mention of the compound), and the names phosphatidyl ethanolamine, phosphatidyl serine and phosphatidyl choline used consistently.

An important feature of the glycerophospholipids is their asymmetry at the central carbon atom of the glycerol. Thus, all naturally occurring phospholipids have the same stereochemical configuration, much in the same way as most natural amino acids are of the '*L*' series. Unfortunately, there has been much disagreement about the convention for naming the stereoisomers. In the past, the members of the natural series have been referred to as *L*–*α*–compounds and represented by the Fischer projection. In this case the glycerol derivative is put into the same category with that glyceraldehyde into which it would be transformed by oxidation, without any alteration or removal of substituents. As the glycerol hydroxyls *can* be distinguished from each other, it is not clear from this nomenclature whether '*α*' refers to the 1– or the 3–hydroxyl of glycerol; for this reason, *L*–3–phosphatidyl choline, for example, would be preferred (see Fig. 1.1). Others have used the equivalent alternative, *D*–1–phosphatidyl choline because it more closely followed the I.U.P.A.C.* convention that a compound be put into

* IUPAC: International Union of Pure and Applied Chemistry.

that optical series which gives its substituent the lower number. Another reason is that the natural diglycerides derived from phospholipids have almost universally been called D–αβ–diglycerides.

Recently, an international committee (the I.U.P.A.C.–I.U.B.* Commission on Biochemical Nomenclature) has recommended the abolition of the DL terminology and provided rules for the unambiguous numbering of the glycerol carbon atoms. Under this system phosphatidyl cholines become 1,2-diacyl-sn-glycero-3-phosphoryl cholines, or the shorter generic term would be: 3-sn-phosphatidyl choline. The letters sn stand for stereochemical numbering and indicate that this system is being used. The stereochemical numbering system is too cumbersome to use with regularity in a book of this type and therefore, normally we shall use the terms 'phosphatidyl choline' etc., but introduce the more precise name when necessary. A brief, but clear introduction to the subject of phospholipid stereochemistry can be found in 'Phospholipids' by Ansell and Hawthorne.

While on the subject of stereochemistry, it is probably worth while to point out that the asymmetry of phospholipids leads to optical activity. Measurement of the specific rotation of a phospholipid is more useful to the organic chemist who is preparing stereochemically pure synthetic lipids than to the biochemist. What has not been generally realized until quite recently is the fact that many triacyl glycerols (triglycerides) also possess

small but measurable rotations and this has provided a useful tool in the study of triglyceride structure.

Another field in which the nomenclature has grown up haphazardly and has recently been revised, is that of the enzymes of lipid metabolism. Quite often the trivial name of an enzyme gives no indication of the reaction catalysed, or there may be confusion between different names for the same enzyme. Nevertheless the Enzyme Commission (EC) nomenclature is very cumbersome for routine use and whenever the trivial name is sufficiently descriptive we will use it in this book but state the EC name at the first mention. Where confusion exists (as in the phospholipases, see chapter 4) or the trivial name is not descriptive, then we shall use the EC name.

Biochemical knowledge has expanded so rapidly in recent years and the subject has become so diverse that it is becoming extremely difficult for a research worker to maintain an interest in all areas of the subject. The result is an increased tendency toward specialization. The student cannot afford to narrow his view to this extent and it is important to avoid regarding any one particular branch of biochemistry such as 'lipids' or 'sugars' as isolated disciplines, bearing no relationship with each other. Of course, all metabolic pathways are interrelated and in this book, we have tried to indicate the way in which reactions involving lipids are intimately connected with the whole functioning of an organism.

Until quite recently, most of the work on lipid metabolism had been done using a limited number of animal species – the rat has been a favourite source of material. Lipid metabolism of bacteria

* IUB: International Union of Biochemistry.

is now in vogue, but higher plants have had very little attention. We have tried not to narrow our attention to a few species but to choose examples from as wide a range as possible from both plant and animal kingdoms.

At the beginning, we stated that this is an introductory book for the student, not a research treatise. For this reason, we have preferred not to quote references in the text for each statement we have made, in the manner of a research paper or a review. At the end of each chapter will be a limited number of references to reviews which cover the appropriate topics in more detail and from which references to original and more specialized papers can be obtained. We also hope that the reader will select sections from different chapters to read in conjunction with each other, rather than reading the book in a 'cover to cover' manner. In a book of this size there is insufficient space for repetition and we hope that the system of cross referencing will be found adequate. As an example of what we mean, we might take the case of *sphingosine*. Although this base is an important constituent of some phospholipids, it will only be briefly mentioned in chapter 4 but treated in more detail in chapter 5.

ANALYTICAL TECHNIQUES

Natural lipids are complex mixtures of chemical species and of permutations of types of fatty acids esterified in each lipid class. This complexity hindered progress in lipid biochemistry because the normal chemical purification procedures of distillation, crystallisation, solvent extraction etc. were inefficient. Liberation, in the sense of ability to study single species, came only with the advent of chromatographic techniques, mainly in the years 1940 to 1960.

First catch your lipid

Before the different lipids can be resolved they must first be separated from all other chemical types present. To do this, advantage is taken of their low water solubility and preference for water immiscible organic solvents, conferred by the long fatty side chains. Since membrane and plasma lipids are normally associated with proteins, solvents having some water solubility and hydrogen bonding ability are necessary to split the lipid-protein complex and even in some cases denature the protein (see chapter 6). The most common extraction solvent is a mixture of chloroform and methanol (2 : 1 by volume). The tissue is best first broken up by use of a homogenizer or an ultrasonic probe, then extracted with the solvent and the organic layer removed. Addition of an equal volume of water will then cause a phase separation and the chloroform layer (containing the lipids) can be separated, washed with water, and dried over an absorbent much as anhydrous sodium sulphate or magnesium perchlorate. After filtering, the solvent can be vacuum evaporated to yield a crude lipid residue that should be protected from oxidation by a blanket of inert gas.

In plant tissue there is a danger of activating the lipases on extraction and so splitting many of the lipids to give free fatty acids. This is best prevented by a prior extraction with *iso*-propanol which inactivates the lipases.

Crude lipid separations can be done by selective extraction

For some experiments it is sufficient to have a crude resolution of phospholipid and neutral lipid, and this can be achieved by extracting the dry lipid with cold dry acetone, when neutral lipids dissolve and most of the phospholipids remain behind. There is, however, still some carry-over of polar lipid into the neutral lipid.

Where only the mixed fatty acids are required the whole tissue or lipid extract can be hydrolysed by dilute aqueous or methanolic potassium hydroxide. The alkaline solution if extracted with light petroleum will yield a non-saponifiable fraction (e.g. sterols) and by extraction with ether after acidification the fatty acids themselves are obtained.

Exploitation of chromatography

Unambiguous separation of lipid classes, molecular species of each class and individual fatty acids can be obtained only by chromatographic techniques. One of the great advantages of chromatography lies in its ability to handle very small amounts of material; indeed the limit is set only by the sensitivity of their detection. Fractional distillation on the other hand has to have enough material to wet the column packing and is difficult with less than about five grams of substance unless carrier distillation is used. Furthermore the high

efficiency of a chromatogram enables molecules to be separated that possess only small differences in structure. The mechanism will be made clear later.

The principles of chromatography are based on distribution between two phases, one moving, the other stationary.

A chromatogram (so-named by its Polish inventor, Tswett, because he used the technique to separate plant pigments) consists of two immiscible phases. One phase is kept stationary either by being held on an inert microporous support or being itself a microporous or particulate adsorbent solid: the other phase is percolated continuously through the stationary phase. The phase pairs that can be used are as follows (the mobile or moving phase being given first), together with their inventors and developers.

1. Liquid–solid (Tswett, 1911, Poland; Kuhn and Lederer, 1931, Germany)
2. Liquid–liquid (Martin and Synge, 1941, U.K.)
3. Liquid–charged gel (Moore and Stein, 1946, U.S.A.; Partridge, U.K.)
4. Liquid–uncharged gel (Boldingh, 1953, The Netherlands; Porath, 1959, Sweden; Determan, 1966, Germany)
5. Gas–solid (Claesson, 1946, Sweden; Phillips, 1949, U.K.; Turkeltaub, 1950, U.S.S.R.)
6. Gas–liquid (James and Martin, 1952, U.K.)

The liquid systems are capable of being used for all substances irrespective of molecular weight, whereas gas–liquid and gas–solid chromatograms can be used only for substances of molecular weight of up to around 700–800.

If we take any single substance and mix it with any of these phase pairs, it will distribute itself between the two phases, the ratio of the concentrations in the two phases (at equilibrium) being known as the partition coefficient.

The partition coefficient is a physical constant dependent on the nature and magnitude of solute-solvent interactions in the two phases. Let us consider two substances A and B and imagine that at equilibrium, substance A distributes itself between the two phases so that 90% is in the stationary phase and 10% in the moving phase. Substance B, however, distributes so that 10% is in the stationary phase and 90% in the moving phase. Then a mixture of A and B dissolved in a small volume of moving phase and applied to the chromatogram will begin to separate when the moving phase is added and washes them through the system. They will distribute themselves independently of one another: B will move as a zone at 9/10ths of the velocity of the moving phase and A at 1/10th of the velocity. Clearly the two substances will rapidly move apart, the original rectangular profile of the zones changing to the shape (ideally) of a Gaussian error curve because of diffusion. The substances can either be visualized *directly on the system* by colour sprays or can be *eluted* as pure components

The two phases can be arranged in a variety of ways

There are basically only two types of chromatogram geometry.

(1) *The column* consists of a metal, glass or even plastic tube with a ratio of length to diameter of at least 10:1 and packed with either an adsorbent solid (silica gel, alumina etc.) or an inert solid, such as kieselguhr, of large surface area, that can hold by surface tension, a liquid as one member of the phase pair. Gas chromatograms can afford a much greater ratio of length to diameter because of their inherent lower resistance to flow than can liquid columns. The surface of kieselguhr normally wets with the more aqueous of a phase pair but this can be reversed by making its surface water-repellent by chemical treatment so that it will hold the least polar of the liquids, hence the term *reversed phase chromatography*. This technique was introduced by Howard and Martin in the U.K. specifically to separate long chain fatty acids, since their partition coefficients were so much in favour of the least polar of a phase pair, that if this were the moving phase, they moved too quickly through the column and hence showed inadequate resolution.

(2) *The plate or strip* consists of the stationary phase support arranged as a flat surface. Mixtures can be spotted and dried on the surface and when the bottom of the plate is immersed in the moving phase in a closed vessel, capillarity ensures that the liquid will move through the porous material (paper, or a porous solid held to a glass or metal surface). The fact that both phases exist as relatively thin films means that the solutes have only short distances to move as they pass from phase to phase. Very refined separations can thus be

Table 1.1. Types of chromatogram used for lipid separations

Phase pair	Geometry	Stationary phase	Moving phase	Separation
Liquid/liquid	Column	Hydrocarbon	Aqueous–organic solvents	Long chain fatty acids
	Column	Aqueous buffer	Ether	Short chain fatty acids
	Thin layer or paper strip	Hydrocarbon or silicone	Aqueous acetone	Long chain fatty acids Neutral lipids
Liquid/uncharged gel	Column	Swollen gel (lipophilic sephadex)	Aqueous organic solvents	Neutral lipids
Liquid/ionized gel	Column	Swollen gel (diethylaminocellulose)	Chloroform, chloroform—methanol, chloroform—methanol—ammonia	Separation of neutral lipids from basic and acidic charged lipids (phospholipids/sulpho-lipids).
Liquid/solid	Column	Silica gel or modified silica gel (e.g. silver nitrate-impregnated)	Range of solvents	all types of lipids
	Paper strip	Silica gel or modified silica gel (e.g. silver nitrate-impregnated)	Range of solvents	all types of lipids
	Thin layer	Silica gel or modified silica gel (e.g. silver nitrate-impregnated)	Range of solvents	all types of lipids
Gas/liquid	Column	High molecular weight hydrocarbons, silicone greases, polyesters	Permanent gas	Fatty acids of all chain lengths, glycerides of all types.

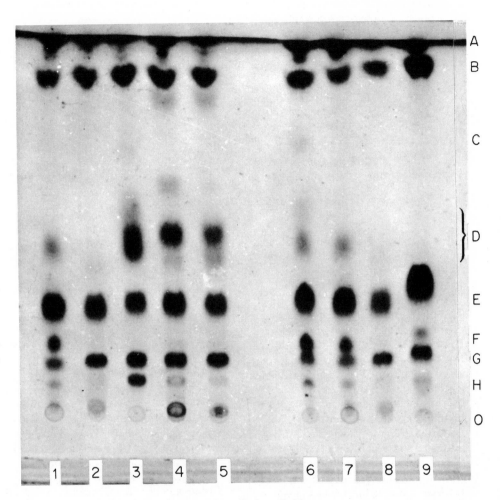

Fig. 1.2. Thin layer chromatogram of plant lipids.
Developed in chloroform-methanol-acetic acid-water (85:15:10:4, by vol.), on silica gel G. Compounds detected by spraying with 50% sulphuric acid and charring. The source of the lipids is as follows:

1. *Chlorella vulgaris.*
2. *Anacystis nidulans.*
3 – 5. Nitrogen-fixing blue-green algae.
6. Spinach leaves.
7. *Chlorella vulgaris.*
8. *Anacystis nidulans.*
9. Spinach chloroplasts.

Fig. 1.2. continued

Identification of the lipids:

A. Neutral lipids.
B. Monogalactosyl diglyceride.
C. Sterol glycoside.
D. In 1, 6 and 7, Phosphatidyl ethanolamine.
   In 3 − 5, A mixture of glycosides of a fatty alcohol.
E. Digalactosyl diglyceride + phosphatidyl glycerol.
F. Phosphatidyl choline.
G. Sulphoquinovosyl diglyceride.
H. In 1, 6 and 7 Phosphatidyl inositol.
   In remainder: unknown.
O. Origin of application.

The R_f of, for example, spot B relative to spot C is the ratio of the distances of the centres of spots B and C from the origin O. This depends only on the solvent, the nature of B and C, and the temperature.

obtained rapidly. The thin layer chromatogram is particularly useful for lipid separations. The plates, which are usually of glass, are prepared in the following manner:

The plate, carefully cleaned so as to be grease-free, is laid on a horizontal surface and is coated evenly with an aqueous slurry of a suitable powdered adsorbent, usually silica gel. The adsorbent sometimes has a binder, such as calcium sulphate, added so as to increase the mechanical strength of the layer. The plate is then heated in an oven at a fixed temperature and for a fixed time so as to 'activate' the adsorbent.

'Activation' is the process of removing water from the gel: the lower the water content the higher the adsorption. The plates can then be stored either in airtight containers or at a fixed relative humidity.

In all types of chromatogram, complexing agents can be added to one or other phase to change the distribution coefficient in favour of that phase by complexing with specific chemical types e.g. borate ions for complexing with cis-hydroxyls, silver ions for complexing with double bonds (see chapter 2, Figs. 2−23, 2−24).

In general, compounds are not eluted from flat plate chromatograms: instead, the development (i.e. the movement of the liquid phase) is stopped when the front has reached the end of the strip, the strip is dried to remove solvent and the position of the zones revealed by spraying. The sprays can be of a destructive type such as dilute sulphuric acid followed by heating (this produces black spots by carbonisation where there is an organic material) (Fig. 1.2) or a non-destructive type such as dichloro- or dibromofluoresceins that show a changed fluorescence where there is a zone. In the latter case the compounds can be recovered by scraping the adsorbent from the plate where the spray indicates a zone to be present, followed by extraction with a suitable solvent. Under standard conditions it will be found that a given substance will move relative to a standard substance to a constant ratio or *relative* R_F (See Fig. 1.2). This *relative* R_F is a useful confirmation of structure of an unknown substance but it is unwise to use it as an absolute indicator.

Only infrequently can every component of a complex mixture be resolved with one solvent system. However, by using two-dimensional development, i.e. first by one solvent system in one direction, then after drying the plate by running a different solvent at right angles, refined separations can be achieved. This is demonstrated in Fig. 1.3 for separation of plant lipids. The volatile solvent (containing ammonia) should be run first as it is difficult to remove the last traces of acetic acid. This lessens the effectiveness of the basic solvent and results in poor separations.

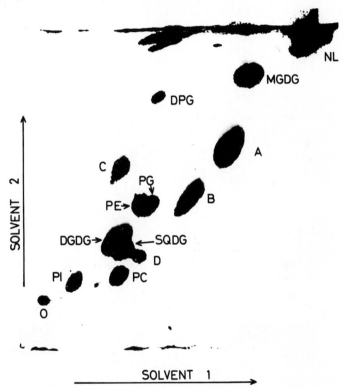

Fig. 1.3. Separation of a complex mixture of plant lipids by two-dimensional thin layer chromatography. Solvent 1: Chloroform-methanol-7N Ammonia, (65:25:4, by vol.). Solvent 2: Chloroform-methanol-acetic acid-water (170:15:15:2, by vol.). O: origin of application; PI: phosphatidyl inositol; PC: phosphatidyl choline; DGDG: digalactosyl diglyceride; SQDG: sulphoquinovosyl diglyceride; PE: phosphatidyl ethanolamine; PG: phosphatidyl glycerol; DPG: diphosphatidyl glycerol; MGDG: monogalactosyl diglyceride; NL: neutral lipids; A, B, C, and D: unknown. Compounds detected by spraying with 50% H_2SO_4 and charring.

Where the zones are radioactive they can be detected by passing the plate under a windowless proportional counter masked by a thin slit. A typical trace is shown in Fig. 1.4. The pure substances can then be isolated by extraction after removing the appropriate area of adsorbent from the plate.

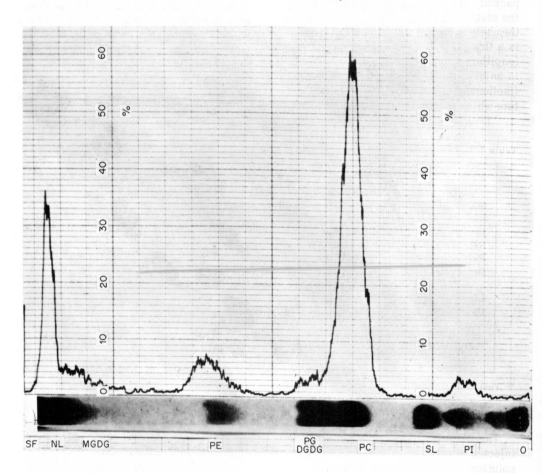

Fig. 1.4. Detection of radioactive substances on a thin layer chromatogram by an automatically recording scanner. The lipids had been extracted from *Chlorella vulgaris* cells grown in the presence of $[1-^{14}C]$oleic acid. Stationary phase: silica gel G, developing solvent: chloroform-methanol-acetic acid-water (85:15:10:4, by vol.). NL: neutral lipids; MGDG: monogalactosyl diglyceride; PE: phosphatidyl ethanolamine; PG: phosphatidyl glycerol; DGDG: digalactosyl diglyceride; PC: phosphatidyl choline; PI: phosphatidyl inositol; SL: sulphoquinovosyl diglyceride. SF: solvent front; O: origin of application.

Gas—liquid chromatography

In this technique the moving phase is a permanent gas and the columns are either packed with *Celite*, on whose surface is the stationary liquid phase, or else are themselves narrow tubes on whose wall is a thin layer of the stationary phase (capillary columns). The column is held in an oven either at a fixed temperature (isothermal operation) or at a temperature which increases during the separation (temperature programming). The latter technique is particularly useful when the mixture to be separated contains components of a wide range of molecular weights. At the end of the column is the vapour detector which responds to the presence of a vapour in the gas stream so that the detector output is proportional either to the instananeous mass of material or its concentration. The two major detectors are compared in sensitivity in Table 1.2. A schematic diagram

Table 1.2. Comparison of the characteristics of two common gas chromatogram detectors

Characteristic	Flame ionization detector	Electron capture detector
Standing current (amps)	$1 \cdot 5 \times 10^{-11}$	3×10^{-9}
Noise level (amps)	5×10^{-15}	2×10^{-12}
Ionization efficiency (%)	$2 \cdot 3 \times 10^{-3}$	—
Relative sensitivity to lead alkyls	1	130
Relative sensitivity to steroids	1	100
Detection limits: benzene (g/ml)	$1 \cdot 0 \times 10^{-12}$	—
aldrin (g)	—	$0 \cdot 5 \times 10^{-12}$
Linear dynamic range	5×10^6	700—1200*
Usage	All types of component	Substances containing electron capture groups eg. halogenated compounds, unsaturated compounds etc.

* This is very dependent on operating conditions and on the substances detected.

of the apparatus is shown in Fig. 1.5.

The mixture to be separated is usually injected in pure form, if liquid, or in solution, if solid, through a flexible septum at the top of the column, so that the material is heated as rapidly as possible to the column temperature. The detector output is usually displayed as a trace on a potentiometric recorder (Fig. 1.6) or as a printed-out record of the area under each peak. At a fixed temperature, with a fixed amount of stationary phase, a fixed pressure drop across the column and a fixed rate of gas flow from the column, a given substance will emerge from the column at a constant time. However since there is a slow loss of stationary phase it is unwise to use *emergence time* or retention volume (the volume of gas required to sweep it from the column) for positive identification of a zone. A better constant is the *relative*

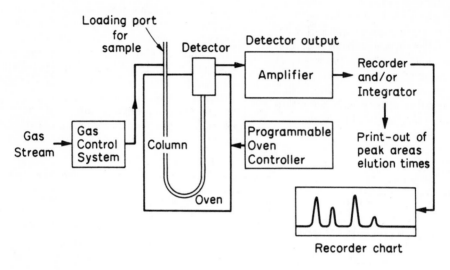

Fig. 1.5. Schematic diagram of gas chromatogram.

retention volume, i.e. the time of emergence relative to a master compound run at the same time. Such a constant can be obtained with two or more chemically distinct types of stationary phase and this can often afford positive identification (see section on Separation of Fatty Acids in Chapter 2). Members of a homologous series of compounds, e.g. the fatty acids, are simply related in emergence times. Thus, if a straight chain fatty acid of 14 carbon atoms emerges after 1 min at 186° on a stationary phase of a high vacuum grease (Apiezon L), then the 15-carbon acid will emerge at 2·3 min, the 16-carbon at $2·3^2$ min, the 17-carbon at $2·3^3$ min etc. By plotting the logarithm of the absolute retention time (or better, relative retention time) against the chain length, a straight line results (Fig. 1.7).

This enables the retention time of a higher or lower homologue to be predicted from knowledge of the retention time of a few members of the series.

Since substances move through the column as vapours in the gas stream, the gas–liquid chromatogram can be used only for substances sufficiently volatile to exert a partial pressure of a mm or so when dissolved in the stationary phase. The temperature instability of stationary phases and of solutes is such that the gas chromatogram is limited to compounds of molecular weight of up to 800 or so. The boiling point of a substance is no guide either to its ability to run on a gas chromatogram or to its retention volume, since the boiling point of a pure substance can be elevated by virtue of strong inter-action forces such as hydrogen bonding

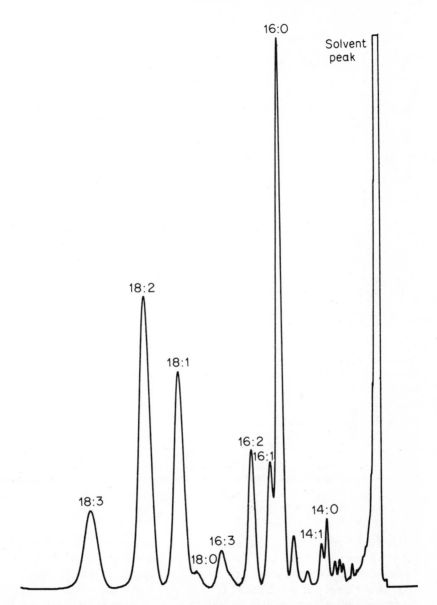

Fig. 1.6. Separation of long chain fatty acid methyl esters by gas-liquid chromatography. A gas-liquid chromatogram of the fatty acid methyl esters derived from the lipids of *Chlorella vulgaris*. Stationary phase: FFAP; temperature: 210° (isothermal); detector: flame ionization. Vertical axis: recorder deflection; horizontal axis: time.

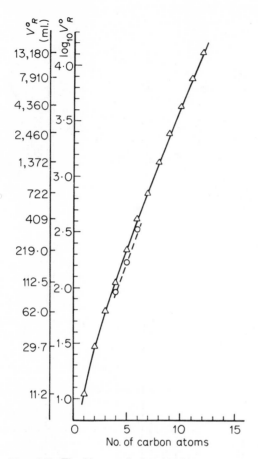

Fig. 1.7. The linear relationship between corrected retention volume and chain length for short chain fatty acids at 137°. Stationary phase: silicone DC 550 containing stearic acid.

— Δ — straight chain acid

— o — *iso*-branched chain acid.

in the liquid state. If the stationary phase of the gas column is a non-polar hydrocarbon or silicone grease then the time of emergence of a compound is determined more by its molecular weight and to a secondary extent by its molecular configuration, than by its boiling point. (see Fig. 1.8 for the relationship between b.pt. and relative retention volume for different types of hydrocarbon). Suitable choice of the stationary phase will allow selective interaction between specific groups and the stationary phase e.g. —OH groups with —O—, or double bonds with ester groups, so that certain molecular types

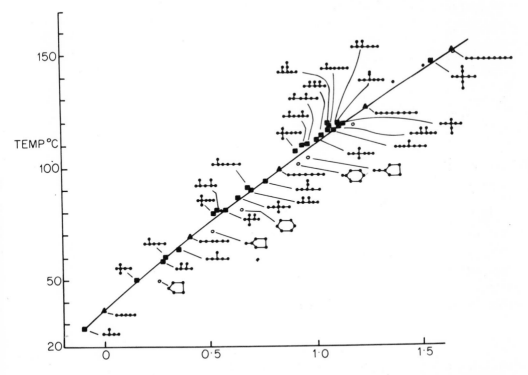

Fig. 1.8. Relationship between logarithm of retention volume relative to *n*-pentane and boiling
point for straight and branched chain and cyclic hydrocarbons. Stationary phase:
n-octadecane; temperature: 65°.
The cyclic hydrocarbons all have greater retention volumes than would be predicted
by comparison of the boiling points of different types.

can be selectively retarded or advanced
(see chapter 2.).

The gas-liquid chromatogram is particularly useful for fatty acids, their esters,
monoglycerides, diglycerides and tri-
clycerides. In general, compounds with
more than one hydroxyl group are best
run as derivatives such as trimethyl
silyl ethers so that the hydroxyl groups
are masked. Otherwise adsorption occurs
and the zones are unsymmetrical (tailing)
and hence can overlap.

The gas-liquid chromatogram can be
made even more useful by having a radioactivity detector in series with a mass
detector. In this way, the ratio of the two
peak areas produced by a single substance
can give directly the *specific activity*.
Figure 1.9 shows a typical record obtained
from such a radiochemical gas chromatogram.

Liquid—gel system

Here the stationary phase is a solvent-

MASS TRACE

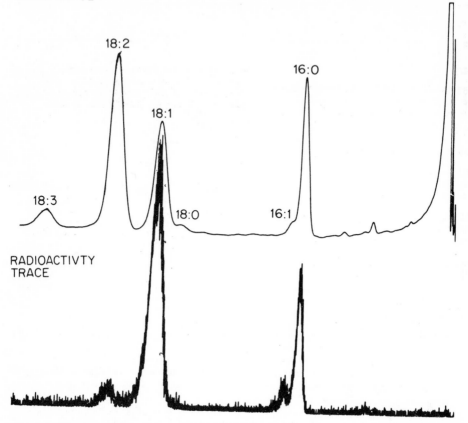

Fig. 1.9. The separation of fatty acids by GLC. A gas-liquid radiochromatogram of fatty acid methyl esters derived from the lipids of flax seeds incubated in $[2-^{14}C]$ sodium acetate, upper trace: mass; lower trace: radioactivity.

swollen gel which may be either charged (ion exchangers) or uncharged. In the latter case systems such as *Lipophilic Sephadex* can give useful separation of some lipid classes. The ion exchangers, e.g. the basic or acidic modified celluloses or Sephadexes are particularly useful for separation of uncharged from basic or acidic lipids.

A summary of the types of chromatogram and their application is given in Table 1.1. In each later chapter concerned with a given lipid class appropriate separation techniques are given in more detail.

BIBLIOGRAPHY

Chemistry

1. GUNSTONE F.D. (1967). *An Introduction to the Chemistry and Biochemistry of Fatty Acids and their Glycerides*. Chapman & Hall, London.

Nomenclature, Stereochemistry

2. ANSELL G.B. and HAWTHORNE J.N., (1964), *Phospholipids*, Elsevier, Amsterdam.

3. IUPAC-IUB Commission on Biochemical Nomenclature: The Nomenclature of Lipids. *Biochem. J.* 105, 897 (1967).

4. Enzyme Nomenclature: Recommendations (1964) of the International Union of Biochemistry on the Nomenclature and Classification of Enzymes together with their units and the Symbols of Enzyme Kinetics. Elsevier, Amsterdam, (1965).

Separation Techniques

5. JAMES A.T. and MORRIS L.J. (eds.) (1964) *New Biochemical Separations*, D. van Nostrand, London. There are several articles in this collection dealing with various aspects of lipid chromatography.

6. MARINETTI G.V., (1967). *Lipid Chromatographic Analysis* Vol. 1, Marcel Dekker, New York.

2 Fatty acids

It is customary in review articles on fatty acids to deplore the general refusal to adopt a single systematic chemical nomenclature and then for the author himself to use a mixture of systematic and trivial names. We shall be no exception but feel that the least we can do is to provide clues as to the solving of the various codes used by workers in all the fields of lipid research. The poor student was still, therefore, have to learn that stearic acid, stearate, $C_{18:0}$ and 18:0 all refer to n-octadecanoic acid, and to commit to memory all the trivial and sometimes tongue-twisting names. The basic chemical nomenclature is based on the hydrocarbon chain having the largest number of carbon atoms, substituents *usually* being defined in position by counting from the carboxyl group as 1 or from carbon atom 2 as the α position. However, in comparing structures of some unsaturated acids it is often convenient to define the double bond position with respect to the terminal methyl group, as acids derived from one another by chain elongation can then be clearly identified. Thus linoleic acid can be referred to either as $(\Delta-)9, 12$-octactecadienoic acid or ω-6,9-octadecadienoic acid.

Analytical data obtained by gas chromatography are usually reported in shorthand by showing the number of carbon atoms followed by a colon and then a figure denoting the number of double bonds. Thus palmitic acid is 16:0 (or $C_{16:0}$); palmitoleic acid is 16:1 or 9—16:1, (or $\Delta^9 C_{16:1}$), hexadecadienoic acid is 16:2 ($C_{16:2}$) and so on. Chain branching or substitution is denoted by a prefix thus: br—16:0 for a branched chain hexadecanoic acid or HO—16:0 as a hydroxy-palmitic acid. This shorthand has the merit of not attributing a more positive identification than the gas chromatogram alone can give.

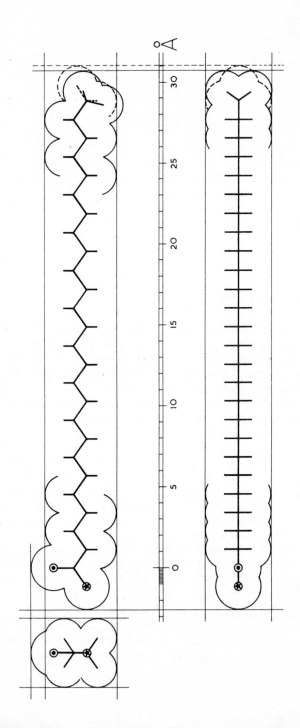

Fig. 2.1. Isometric projection of a long chain saturated acid (24:0 lignoceric acid) after Vandenheuvel.

Straight chain fatty acids

Table 2.1 gives a list of some straight chain acids, their systematic and trivial names, melting point and some data on their occurrence. In Tables 2.1 to 2.5 we shall only list those acids whose biochemistry we shall discuss later in the chapter. More information can be found in References 1–4 at the end of the chapter. In general, fatty acids do not exist as free carboxylic acids because of their marked affinity for many proteins. One result of this is an inhibitory action on most enzymes. Where free acids have been reported as major constituents they are usually artefacts due to cell damage which allows lipases to act on acyl lipids of the tissue. One major exception to this rule lies in the albumin bound fatty acids (ABFA) of mammalian blood. You will also see them referred to as FFA [free fatty acids] or NEFA [non-esterified fatty acids] (see chapter 6).

The unperturbed configuration of a typical saturated chain is shown in Fig. 2.1. It should be remembered that the chain is in continuous thermal motion in living systems, and that the close chain packing studies by means of Langmuir troughs, X-ray diffraction of crystals etc., is not necessarily relevant to membrane properties. Nevertheless the physical properties of the individual fatty acids do affect the physical properties of acyl lipids, one of the most obvious being the melting point. No living organism appears to be capable of operating with lipids that are solid at 37 °C. Having said this, it must be immediately qualified since many purified phospholipids are amorphous solids at room temperature, but it should be remembered that cellular lipids consist of complex mixtures of both phospholipids and neutral (i.e. uncharged) lipids containing a wide range of fatty acids. This complexity usually ensures that the mixture is semi-liquid at 37 °C; where organisms are subjected naturally to a range of temperatures their lipids are usually highly unsaturated and hence of low m.pt.

While most natural fatty acids are even-numbered due to their mode of biosynthesis, it should be noted that odd numbered fatty acids *do* occur naturally. The biosynthesis of both types will be discussed in a later section.

Branched Chain Fatty Acids

The majority of branched chain fatty acids are mono-methyl substituted acids. All have lower m.pts than the corresponding straight chain acids owing to the side groups preventing close approach of the chains. There are two distinct series, the *iso*-series having the terminal group:

$$CH_3 \cdot CH-$$
$$|$$
$$CH_3$$

and the *anteiso*-series having as terminal group:

$$CH_3 \cdot CH_2 \cdot CH-$$
$$|$$
$$CH_3$$

Major sources of both series are wool fat, butter fat, and bacteria.

The highly branched acids isolated from various strains of the *tubercle bacillus* attracted attention because of their ability to induce inflammatory

Table 2.1. Some naturally occurring straight chain saturated acids

No. of carbon atoms	Systematic name	Common name	M. pt. (°C)	Occurrence
2	n-Ethanoic	Acetic	16.7	As alcohol acetates in many plants, and in some plant triglycerides. At low levels widespread as salt or thiolester. At higher levels in the rumen as salt.
3	n-propanoic	propionic	−22.0	At high levels in the rumen
4	n-Butanoic	Butyric	−7.9	At high levels in the rumen, also in milk fat of ruminants
8	n-Octanoic	Caprylic	16.7	Very minor component of animal and plant fats
10	n-Decanoic	Capric	31.6	Widespread as a minor component
12	n-Dodecanoic	Lauric	44.2	Widely distributed, a major component of some seed fats
14	n-Tetradecanoic	Myristic	54.1	Widespread occasionally as a major component
16	n-Hexadecanoic	Palmitic	62.7	Widespread usually as a major component. One of the commonest fatty acids in both plants and animals
18	n-Octadecanoic	Stearic	69.6	Widespread usually as a major component
20	n-Eicosanoic	Arachidic	75.4	Widespread minor component, occasionally a major component
22	n-Docosanoic	Behenic	80.0	Fairly widespread as minor component in seed fat triglycerides
24	n-Tetracosanoic	Lignoceric	84.2	Fairly widespread as minor component in seed fat triglycerides
26	n-Hexacosanoic	Cerotic	57.7	Widespread as component of plant and insect waxes
28	n-Octacosanoic	Montanic	90.9	Major component of some plant waxes

Table 2.2. Some natrually occurring mono- and di-unsaturated fatty acids

No. of carbon atoms	Systematic name	Common name	M.pt. (°C)	Occurrence
Monoenoic acids				
16	*trans*-3-hexadecenoic			Plant leaves; *Chlorella vulgaris*; specifically as component of phosphatidyl glycerol
	cis-5-hexadecenoic			Ice plant, Bacilli
	cis-7-hexadecenoic			Algae, higher plants, bacteria
	cis-9-hexadecenoic	palmitoleic		Widespread: animals, plants, micro-organisms. Major component in some seed oils
18	*cis*-11-hexadecenoic	palmitvaccenic		Some seed oils
	cis-9-octadecenoic	oleic	10.5	Probably most common fatty acid in plants and animals.
	cis-11-octadecenoic	vaccenic	13.0	Also found in micro-organisms *E.coli* and other bacteria
Dienoic acids				
16	*cis*, *cis*-7,10-hexadecadienoic			Mainly plants and algae
	cis, *cis*-9,12-hexadecadienoic			minor component
18	*cis*, *cis*-6,9-octadecadienoic			Minor component in animals
	cis, *cis*-9,12-octadecadienoic	linoleic	–5.0	Major component in plant lipids. In animals it is derived only from dietary vegetables, and plant and marine oils

Table 2.3. Some naturally occurring tri, tetra, penta and hexaenoic fatty acids

No. of carbon atoms	Systematic name	Common name	M.pt. (°C)	Occurrence
Trienoic acids (methylene interrupted)				
16	all-*cis*-7,10,13-hexadecatrienoic			Higher plants and algae
18	all-*cis*-6,9,12-octadecatrienoic	γ-linolenic		Minor component in animals and some algae. Important constituent of some plants
	all-*cis*-9,12,15-octadecatrienoic	α-linolenic	−11	Higher plants and algae especially as component of galactosyl diglycerides
Trienoic acids (conjugated)				
18	*cis*-9, *trans*-11, *trans*-13-octadecatrienoic	α-eleostearic	44	Some seed oils especially Tung oil
Tetraenoic acids				
16	all-*cis*-4,7,10,13-Hexadecatetraenoic			*Euglena gracilis*
20	all-*cis*-5,8,11,14-Eicosatetraenoic	arachidonic	−49.5	A major component of animal lipids and some algae, especially as component of phospholipids
Pentaenoic acids				
22	all-*cis*-7,10,13,16,19 docosapentaenoic	clupanodonic		Animals, especially as phospholipid component Abundant in fish
Hexaenoic acids				
22	all-*cis*-4,7,10,13,16,19 docosahexaenoic			Animals, especially as phospholipid component Abundant in fish

Table 2.4. Structural relationships of some unsaturated acids

Carboxyl-9 series	
$CH_3(CH_2)_7 CH=CH(CH_2)_7 CO_2H$	Oleic
$CH_3(CH_2)_4 CH=CH \; CH_2CH=CH(CH_2)_7 CO_2H$	Linoleic
$CH_3CH_2CH=CH\cdot CH_2CH=CH\cdot CH_2CH=CH(CH_2)_7 CO_2H$	α-Linolenic
ω-6 series	
$CH_3(CH_2)_4CH=CH\cdot CH_2CH=CH(CH_2)_7 CO_2H$	Linoleic
$CH_3(CH_2)_4CH=CH\cdot CH_2CH=CH\cdot CH_2CH=CH(CH_2)_4 CO_2H$	γ-Linolenic
$CH_3(CH_2)_4CH=CH\cdot CH_2CH=CH\cdot CH_2CH=CH(CH_2)_6 CO_2H$	C_{20} trienoic acid
$CH_3(CH_2)_4CH=CH\cdot CH_2CH=CH\cdot CH_2CH=CH\cdot CH_2CH=CH(CH_2)_3 CO_2H$	Arachidonic acid

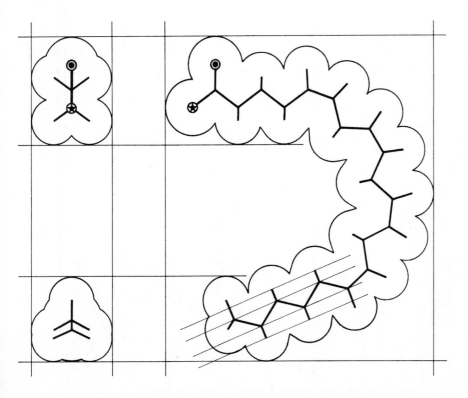

Fig. 2.2. Isometric projection of arachidonic acid (5,8,11,14–20:4) after Vandenheuvel.

Table 2.5. **Structural relationships of some substituted stearic acids**

Structure	Common name	Systematic name
$CH_3(CH_2)_7CH_2-CH_2(CH_2)_7CO_2H$	Stearic	Octadecanoic
$CH_3(CH_2)_7CH=CH(CH_2)_7CO_2H$	Oleic	cis-9-octadecenoic
$CH_3(CH_2)_5CH \cdot CH_2CH=CH(CH_2)_7CO_2H$ $\quad$ OH	Ricinoleic	12-hydroxy-cis-9-octadecenoic
$CH_3(CH_2)_7CH-CH \cdot (CH_2)_7CO_2H$ $\quad$ CH_2	Dihydrosterculic	8-[2-octyl cyclopropanyl]octanoic
$CH_3(CH_2)_7C=C \cdot (CH_2)_7CO_2H$ $\quad$ CH_2	Sterculic	8-[2-octyl cyclopropenyl]octanoic
$CH_3(CH_2)_7CH-CH_2(CH_2)_7CO_2H$ $\quad$ CH_3	10-Methylstearic	10-methyl octadecanoic
$CH_3(CH_2)_7CH-CH(CH_2)_7CO_2H$ $\quad$ O	cis-9,10-epoxystearic	cis-9,10-epoxy octadecanoic
$CH_3(CH_2)_7CH-CH(CH_2)_7CO_2H$ $\quad$ OH OH	9,10-dihydroxystearic	9,10-dihydroxy octadecanoic
$CH_3(CH_2)_7CH-CH_2(CH_2)_7CO_2H$ $\quad$ OH	10-Hydroxystearic	10-hydroxy octadecanoic

reactions similar to the bacillus itself. Their biosynthetic pathway is quite different from that of the *iso* and *ante-iso* acids will be seen on pages 41–43.

Unsaturated fatty acids

(a) *Monoenoic acids.* The more sophisticated separation and identification techniques introduced in the last fifteen years have enormously widened the range of defined naturally occurring acids particularly unsaturated acids. In the monoenoic acid series the major components have a *cis* configuration about the double bond. *Trans*-isomers are rare but do exist, one of the most interesting being *trans*-3-hexadecenoic acid, a major fatty acid esterified in phosphatidyl glycerol of some photosynthetic organelles.

The *cis* double bond introduces a kink in the molecule (Fig. 2.2) so that the molecular shape is clearly distinct from that of the saturated acids. This is not true for *trans* acids: elaidic acid, for example, has physical properties similar to stearic acid and is metabolized more like a saturated acid than an unsaturated acid so far as positional specificity of insertion into acyl lipids is concerned. The *cis* forms are thermodynamically less stable than the *trans* forms and have lower melting points than the corresponding saturated acids for much the same reason as the branched chain acids. Table 2.2 lists some naturally occurring monoenoic acids. Although most natural fatty acids have an *ethylenic double bond,* a few are known to have one or more *acetylenic* bonds. Little is known about their biosynthesis.

(b) *Dienoic acids.* All dienoic acids are derived from monoenoic acids, the position of the second double bond being a function of the biochemical system. Thus mammals have 'desaturase' enzymes capable of removing hydrogen atoms only from carbon atoms between the existing double bond and the carboxyl group. Higher plants on the other hand carry out the same reaction between the existing double bond and the terminal methyl group. The two series of polyunsaturated acids produced by such sequential desaturations will become more obvious when the more highly unsaturated acids are considered (see Table 2.4). Table 2.2 lists some common diunsaturated acids.

Higher unsaturated acids

Table 2.3 gives some data on a few tri, tetra, penta, and hexaenoic acids. Metabolically they are derived from dienoic acids by chain elongation and further desaturation. The hindrance of free rotation brought about by the double bonds can give rise to bent configurations as shown in Fig. 2.2.

Cyclic fatty acids

These are more limited in type than any other group. The ring structures are either cyclopropyl or cyclopentyl; even numbered rings have not been identified as yet. Many micro-organisms produce cyclopropane and cyclopropene acids and the latter also occur in seed oils of certain families. The ones whose biochemistry we shall discuss will be the following: *Lactobacillic* is *cis*-11,12-methylenoctadecanoic acid and is related to *Vaccenic* acid:

$$CH_3(CH_2)_5 \overset{\overset{\displaystyle CH_2}{\diagup \diagdown}}{CH-CH} (CH_2)_9 CO_2H$$

The structures of the cyclopropene acid sterculic, and the cyclopropane acid, dihydrosterculic are shown in Table 2.5. *Chaulmoogric acid* is 13-(2-cyclopentenyl)-tridecanoic acid, found in *Chaulmoogra* seed oil.

Oxy acids

A great range of keto, hydroxy, and epoxy acids have been identified in recent years particularly in association with conjugated unsaturated acids. Table 2.5 indicates those oxy acids, related to stearic or oleic acids, whose metabolism has been studied.

Conjugated unsaturated fatty acids

The major polyunsaturated acids all contain *cis* methylene interrupted sequences and for some years it was thought that most conjugated systems were artefacts of isolation. However, many such acids have now been firmly identified, such as α-*eleostearic* acid (Table 2.3).

Fatty aldehydes and fatty alcohols

Many tissues contain appreciable amounts of fatty alcohols or aldehydes whose chain length and double bond patterns reflect those of the fatty acids themselves, suggesting that they are derived from fatty acids. This is of significance when considering the biosynthesis of vinyl ethers, or plasmalogens (chapter 4) or alkyl ethers (chapter 3), whose acyl chains are derived from fatty alcohols and aldehydes.

Some properties of the fatty acids

The short chain fatty acids (i.e. of chain length up to C_8) are all water-soluble although in solution they are *associated* and do not exist as single molecules. The longer chain acids are much less soluble because of the size of the hydrocarbon chain. The alkali metal salts are the familiar soaps and are very powerful detergents with unusual solubility properties. These unusual physical properties are due to the balance between the long chains, unable to fit into water structure and preferring to interact with one another, and the water soluble charged carboxyl group.

McBain and Salmon many years ago introduced the concept of *micelles*, 'heavily hydrated charged aggregates', based on their demonstration that a 1·0N 'solution' of potassium stearate has an osmotic activity equivalent to only a 0·42N solution of an undissociated salt and equivalent conductivity of 67% of that of the same concentration of potassium acetate. The most striking evidence for the formation of micelles in aqueous solutions of lipids lies in extremely rapid changes in physical properties over a limited range of concentration, the point of change being known as the 'critical micelle concentration' or CMC and exemplifies the great tendency of lipids to self-associate to stable aggregates rather than remain as single molecules. The 'CMC' is not a fixed concentration but a small range of concentration and is markedly affected by the presence of other ions, neutral molecules etc.

Fatty acids are easily extracted from solution or suspension by lowering the pH to form the uncharged carboxyl group and extracting with a non-polar solvent such as light petroleum. The saturated acids are very stable but unsaturated acids are susceptible to oxidative attack, the more double bonds the greater the susceptibility. Both free acids and esters are best kept in dilute solution in light petroleum, in the dark (to prevent photochemical oxidation particularly in the presence of chlorophyll), at low temperatures. Lipid samples should never by handled in pure or concentrated form without a blanket of inert gas.

THE BIOCHEMISTRY OF FATTY ACIDS

The activation of fatty acids

For most of the metabolic reactions in which fatty acids take part, whether they be anabolic (synthetic) or catabolic (degradative), thermodynamic considerations dictate that the acids be 'activated'. The active form is usually the thiol ester of the fatty acid with the complex nucleotide, *Coenzyme A* (CoA) or the small protein known as *acyl carrier protein* (ACP) [see Fig. 2.3]. By this means, two important things are achieved. First, the molecule contains a 'high-energy' thiol ester bond between the carboxyl function of the fatty acid and the –SH group of the 4-phosphopantetheine moiety of the coenzyme. Second, the acyl chains are rendered water-soluble.

In living cells, the activation of fatty acids may be catalysed by the enzyme *thiokinase* (*Acid: CoA ligase* [AMP] *or acyl-CoA synthetase*). Altogether three enzymes have been discovered with this activity, each acting at different parts of the chain length spectrum, but overlapping slightly in their specificities. *Acetic thiokinase* (EC 6.2.1.1) activates acetate and propionate; the *general thiokinase* (EC 6.2.1.2) activates acids with chain lengths of four to twelve carbons, but also phenyl substituted, branched chain, $\alpha\beta$– and $\beta\gamma$–unsaturated and β–hydroxy but not β–keto acids;

finally, the *long chain fatty acid thiokinase* (EC 6.2.1.3) activates fatty acids from C_8 to C_{18}.

Light was thrown on the ATP requirement by Berg, who synthesized acetyl-AMP and showed that the following reactions were catalysed by acetic thiokinase:

$$\text{acetyl}-\text{AMP} + \text{CoASH} \longrightarrow \text{acetyl}-\text{CoA} + \text{AMP} \quad (1)$$

$$\text{acetyl}-\text{AMP} + \text{PP}_i \xrightarrow{\text{Mg}^{++}} \text{acetate} + \text{ATP} \quad (2)$$

Only reaction (2) requires magnesium, and Berg postulated that the first reaction catalysed by thiokinase was the reverse of reaction (2), leading to the formation of acetyl–AMP and pyrophosphate; the second step of the activation process would be reaction (1). A difficulty in accepting this mechanism was the failure to isolate or identify acetyl–AMP in the β–oxidation system. The American biochemist, D.E. Green, at the Enzyme Institute, proposed a series of *enzyme-bound* magnesium chelates which explained both the failure to isolate acetyl–AMP (because it didn't exist as such) and the universal requirement for magnesium in ATP-requiring syntheses.

The enzyme can be assayed by measuring the change in: (*i*) concentration of –SH groups by reaction with nitro-

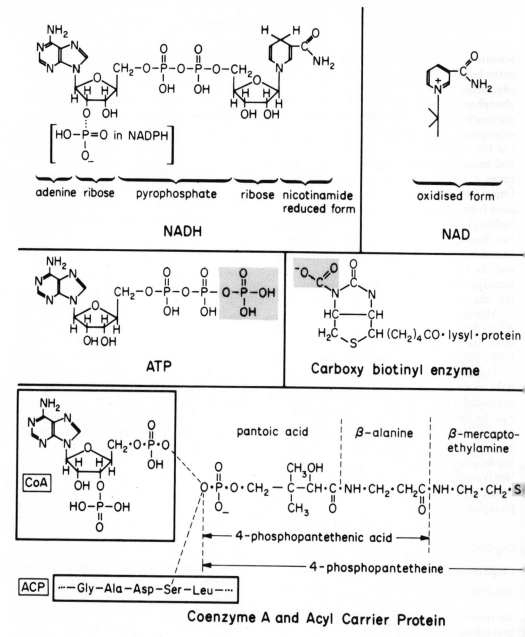

Fig. 2.3. Some coenzymes involved in fatty acid synthesis.

prusside; (ii) the absorption of the thiol ester bond at 232nm; (iii) the concentration of ATP by coupling with the phosphorylation of glucose in the presence of hexokinase; (iv) the concentration of pyrophosphate by reacting the PP_i with a specific pyrophosphatase and measuring inorganic phosphate. Sometimes it is convenient to measure the formation of the hydroxamate of the fatty acid from acyl-CoA in the presence of hydroxylamine. This method has at least two disadvantages: the enzymes are often unstable in the presence of hydroxylamine and the formation of hydroxamates can sometimes be catalysed by lipases in the *absence* of ATP, Mg^{++} and CoA.

Although most activating enzymes are ATP–dependent, some thiokinases have been discovered which are specific for GTP, (Guanosine triphosphate).

The thiokinase mechanism is not the only means by which a fatty acid may be activated. In some micro-organisms, CoA may be transferred from acetyl-CoA to short chain fatty acids (up to C_5) by an enzyme known as *thiophorase*. In these organisms, acetate itself is first 'activated' by phosphorylation and then the acetyl group transferred from acetyl phosphate to CoA:

chemical synthesis of acyl–CoA thiol esters for use as substrates in various enzymic reactions. Several methods are now available, which involve acylation of CoA with acyl chlorides, acyl–N–hydroxysuccinimides or mixed anhydrides of the fatty acid. Of course, acyl–CoA substrates may also be prepared *biochemically by* choosing a suitable activating system.

The biosynthesis of saturated fatty acids

Fatty acids are built up by successive condensations of 2-carbon units.

One of the features emerging from a study of the structure of naturally occurring fatty acids is that most of them have an even number of carbon atoms in the chain. Early speculations that this is because they are built up by successive condensations of two-carbon units were confirmed by the experiments of Rittenberg and Bloch in 1944. They isolated fatty acids from tissues of rats which had been fed acetic acid 'labelled' with ^{13}C in the carboxyl group and deuterium in the methyl group. The two kinds of labelled atoms were located at alternate positions along the chain, showing that the complete

$$CH_3COO^- + ATP \overset{Mg^{++}}{\rightleftharpoons} CH_3COPO_3^- + ADP \qquad (1)$$

$$CH_3COPO_3^- + CoASH \rightleftharpoons CH_3COSCoA + H_2PO_4^- \qquad (2)$$

$$CH_3COSCoA + RCOO^- \rightleftharpoons RCOSCoA + CH_3COO^- \qquad (3)$$

As research on fatty acid synthesis and breakdown has progressed, so it has been necessary to develop methods for the

chain could be derived from acetic acid.

This stimulated interest in the mechanism of chain elongation and when the main details of the β—oxidation pathway (the means by which fatty acids are broken down two carbon atoms at a time, see page 65) were worked out in the early 1950s it was natural for many biochemists to ask the question: can β—oxidation be reversed under certain circumstances to *synthesize* fatty acids instead of breaking them down?

The major pathway for fatty acid synthesis does not involve reversal of β—oxidation. A 2—carbon plus a 3—carbon unit form a four—carbon unit with evolution of CO_2.

The study of fatty acid biosynthesis began in the laboratories of S. Gurin in the U.S.A., who studied liver and G. Popják in London, studying mammary gland. Several discoveries soon indicated that the major biosynthetic route to long chain fatty acids was distinctly different from β—oxidation. In the first place, a pyridine nucleotide was involved; but the reduced form, NADPH, not NAD^+ as in β—oxidation (see Fig. 2.3 and Table 2.6). Second, a requirement for bicarbonate or carbon dioxide was noticed by S.J. Wakil and his colleagues who were studying fatty acid biosynthesis in pigeon liver and by R.O. Brady and H. Klein studying rat liver and yeast respectively.

The study of a biosynthetic pathway usually progresses in well-defined steps: first the overall pathway is demonstrated in the living animal (*in vivo*); then the specific organ where the reaction takes place is located and the reaction demonstrated in the isolated organ or in slices of the organ; then in a cell-free homogenate; next in subcellular components and finally, each step is studied at the level of the isolated, purified enzyme (*in vitro*). Fatty acid biosynthesis was no exception and in pigeon liver, the enzymes of fatty acid synthesis were found to be in the supernatant fraction which could be split into two protein fractions by ammonium sulphate precipitation; both of these fractions added together were necessary to produce fatty acids from acetate, and the coenzyme *biotin* was found to be bound to one of the protein fractions. At that time biotin was already known to be involved as a cofactor in carboxylation reactions and when Wakil discovered soon afterwards that *malonic acid* was an intermediate in fatty acid synthesis, the essential features of the pathway began to emerge. The first step in fatty acid synthesis is the carboxylation of the 2-carbon fragment acetic acid to malonate. The acetate must be in an 'activated form' as its CoA thiol ester (see Fig. 2.3 and p. 31). The carboxylation is catalysed by the enzyme, *acetyl-CoA carboxylase* (Acetyl—CoA: carbon dioxide ligase (ADP) E.C.6.4.1.2) which could be identified with the protein fraction from pigeon liver supernatant to which biotin was bound. One of the first suggestions that biotin was the prosthetic group of acetyl-CoA carboxylase came from the observation that the carboxylation step was inhibited by avidin. This protein was already well known as a potent inhibitor of biotin, for it is the component of raw egg white which can bind to biotin causing a vitamin deficiency known as 'egg white

Table 2.6. Reactions of fatty acid synthesis in *E. coli*

1 Malonyl transacylase	$HOOC \cdot CH_2 \cdot$ ~S·CoA + ACP–SH $\rightleftharpoons$ $HOOC \cdot CH_2 \cdot C$~S–ACP + CoA–SH	Specific for malonate: not a saturated acyl-CoA. Malonyl-S-pantetheine also a substrate. Both enzymes (1 and 2) were characterized by (*i*) the amount of ^{14}C-acetate or malonate transferred to ACP or (*ii*) paper chromatography of acetyl or malonyl hydroxamate. Intermediate is acyl-S-enzyme
2 Acetyl transacylase	CH_3C~S–CoA + ACP–SH $\rightleftharpoons$ $CH_3 \cdot C$~S·ACP + CoA–SH	
3 β-ketoacyl-ACP synthetase	$CH_3 \cdot C$~S·ACP + $HOOC \cdot CH_2 \cdot C$~S·ACP $\rightleftharpoons$ $CH_3 \cdot C \cdot CH_2 \cdot C$~S·ACP + CO_2 + ACP·SH	Assayed by coupling with the next reaction and following NADPH oxidation spectrophotometrically. SH-enzyme; inhibited by iodoacetamide, etc. Protected by preincubation with acetyl-ACP *not* malonyl-ACP.
4 β-ketoacyl-ACP reductase	$CH_3 \cdot C \cdot CH_2 C$~S·ACP + NADPH + H^+ $\rightleftharpoons$ $D(-)$ $CH_3 \cdot CH \cdot CH_2 \cdot C$~S·ACP + $NADP^+$ (OH)	Also reacts with CoA, pantetheine esters but much more slowly. Stereospecific for $D(-)$ isomer

5
β-hydroxyacyl-
ACP dehydrase
	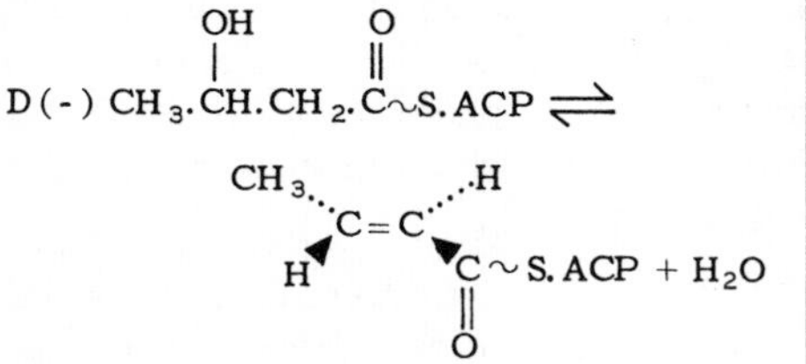 $D(-)$ $CH_3 \cdot CH \cdot CH_2 \cdot C$~S·ACP $\rightleftharpoons$
	(OH)
	
	CH_3 ···· H
	  C=C
	H···· C~S·ACP + H_2O
	     ‖
	     O
		Measured by hydration of crotonyl-ACP accompanied by decrease in absorption at 263 nm. Does not metabolize 'model compounds'. Stereospecific for $D(-)$ isomer.

6
Enoyl-ACP
reductase
	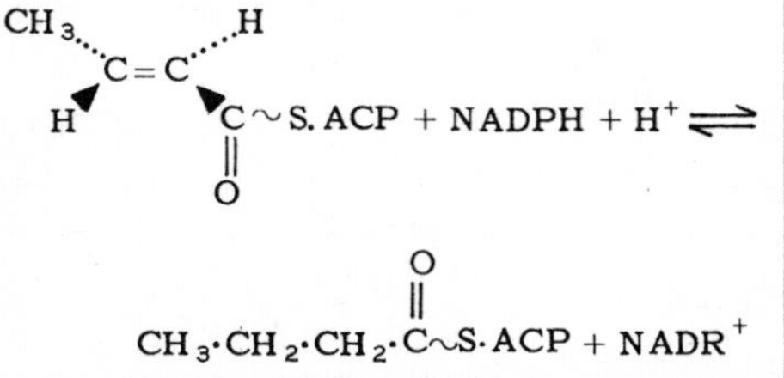 CH_3 ···· H
	  C=C
	H···· C~S·ACP + NADPH + H^+ $\rightleftharpoons$
	     ‖
	     O
	
	$CH_3 \cdot CH_2 \cdot CH_2 \cdot C$~S·ACP + $NADR^+$
		Two enzymes occur; (*i*) NADPH specific, short chain acids preferred; inhibited by iodoacetamide, NEM, PCMB. Specific for ACP esters. (*ii*) NADH specific, long chain acids preferred. Uses CoA or ACP esters. NEM stimulates.

injury'. The *active carbon dioxide* is attached to one of the ureido nitrogens of the biotin ring and is transferred to acetyl–CoA according to the following scheme:

bicarbonate. Bicarbonate (or CO_2) is not itself incorporated into the final long chain fatty acid but the same CO_2 which is used in malonate synthesis is again evolved during the condensation reactions.

$$ATP + \boxed{HCO_3} + \text{biotin-enzyme} \rightleftharpoons \boxed{CO_2} \sim \text{biotin-enzyme} + ADP + P_i$$

$$\boxed{CO_2} \sim \text{biotin-enzyme} + CH_3\overset{O}{\underset{\|}{C}}\cdot S\cdot CoA \rightleftharpoons \text{Biotin-enzyme} + \boxed{OOC} CH_2\overset{O}{\underset{\|}{C}}\cdot S\cdot CoA$$

Acetyl-CoA carboxylase occurs in two forms. If citrate is present, the enzyme consists of aggregates of protein units in the form of long filaments, 70–100Å wide and 4,000Å long. Each aggregate has a total molecular weight of $7\cdot8 \times 10^6$ and is associated with 20 moles of biotin. This is the active form. In the absence

The energy of decarboxylation is used to drive the reaction in the direction of synthesis. The fatty acid grows from the carboxyl group; the two carbon atoms at the methyl end are derived directly from acetate (the 'primer' for the reaction) while the remaining carbon atoms are derived from malonate:

$$\boxed{CH_3.CH_2}.CH_2.CH_2.CH_2.CH_2.CH_2.CH_2.CH_2.CH_2.CH_2.CH_2.CH_2.CH_2.CH_2.COOH$$

from acetyl
—CoA

from malonyl-CoA

of citrate, the aggregates dissociate into 'monomers' of M.W. 410,000, each containing one mole of biotin but having no enzymic activity. Citrate activates fatty acid synthesis in general but the effect can be narrowed down to a specific activation of acetyl-CoA carboxylase. It does not seem to be a property peculiar to citrate or even to tricarboxylic acid cycle intermediates; but rather of polybasic acids in general. We shall discuss the possible role of citrate in the control of fatty acid synthesis at the end of this section.

The building units for fatty acid synthesis, therefore, are acetyl-CoA, and malonyl-CoA which had been formed from acetyl-CoA by carboxylation with

If the 'primer' is an acid with an odd number of carbon atoms such as propionic, then the resulting long chain fatty acid is also an 'odd-chain acid'. If a branched chain primer is used, then a branched chain fatty acid results.

Most acetyl-CoA is produced by mitochondria and special mechanisms must operate to transport the substrates across the mitochondrial membrane to the sites of fatty acid synthesis.

What is the origin of the main building blocks, acetyl-CoA and malonyl-CoA? Most of the cell's acetyl-CoA is derived from the oxidation of pyruvate in mitochondria. A difficulty arises that the mitochondrial membrane is rather im-

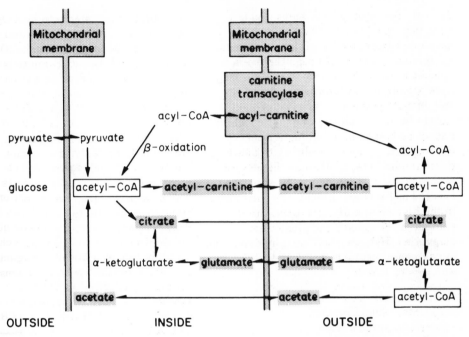

Fig. 2.4. Formation of extramitochondrial acetyl-CoA.

permeable to acetyl-CoA, while the sites for fatty acid synthesis are mainly in the cytoplasm outside the mitochondria. Acetyl-CoA may be *formed* by direct esterification with CoA catalysed by *acetic thiokinase* (see reaction 1 below and page 37) or from citrate by a reaction catalysed by *citrate cleavage enzyme* (2) :

$$\text{Acetate} + \text{ATP} + \text{HS–CoA} \rightleftharpoons$$
$$\text{Acetyl–CoA} + \text{AMP} + \text{PP}_i \quad (1)$$

$$\text{Citrate} + \text{ATP} + \text{HS–CoA} \rightleftharpoons$$
$$\text{Acetyl–CoA} + \text{oxaloacetate} + \text{ADP} + \text{P}_i \quad (2)$$

Acetyl–CoA may be *transported* across the mitochondrial membrane in the four ways illustrated in Fig. 2.4. A fuller discussion of the role of carnitine in fatty acid metabolism will be found on p. 73. The relative importance of these four ways of obtaining acetyl–CoA outside mitochondria is still unclear.

Knowledge of the intermediate steps in fatty acid synthesis has come from studies on yeast, E.coli and pigeon liver.

The 'primer' for fatty acid synthesis is acetyl–CoA and the chain is built up by successive condensations with malonyl-CoA:

$$\text{CH}_3\text{COSCoA} + 7 \boxed{\text{HOOC}} \text{CH}_2\text{COSCoA} + 14\,\text{NADPH} + 14\,\text{H}^+$$
$$\longrightarrow \text{CH}_3(\text{CH}_2)_{14}\text{COOH} + 7 \boxed{\text{CO}_2} + 8\,\text{CoASH} + 14\text{NADP}^+ + 6\text{H}_2\text{O}$$

What are the intermediate steps and how were they elucidated? The answers have come from studies of three different *synthetases*, which, although the overall biochemistry of the reactions is the same, have significant differences. The yeast and animal synthetases consist of tightly bound complexes of enzymes which cannot be broken up into single components that retain their ability to catalyse the individual steps; whereas the enzymes of the *E.coli* synthetase can be separated, purified and studied individually. Although in the historical order of events, the reactions of the yeast synthetase were unravelled first, we shall start by describing the separate reactions catalysed by the enzymes of *E.coli*.

Component enzymes of the E.coli synthetase can be studied separately. Intermediates are bound to an acyl carrier protein.

The Americans, P. Goldman and P.R. Vagelos discovered that the product of the first condensation reaction of fatty acid synthesis in *E.coli* was bound to a protein through a thiol ester linkage:

Acetoacetyl-S-Protein

When they incubated this product with the coenzyme NADPH and a crude protein fraction containing the *synthetase*, they obtained butyryl-S-protein and when malonyl-CoA was included in the reaction mixture, long chain fatty acids were produced. Subsequently, intermediates at all steps of the reaction were shown to be attached to this protein, which was therefore called *Acyl carrier protein* (ACP). It is a small molecular weight protein (about 9,800), has one SH-group per mole of protein and is stable to heat and acid pH. Vagelos and his colleagues worked out the structure of the prosthetic group to which the acyl moieties were attached, in the following way. They made acyl-ACP derivatives in which the acyl group was 'labelled' with radioactive carbon atoms (such as 2-^{14}C-malonyl-ACP) and then hydrolysed the acyl protein with proteolytic enzymes. This yielded small radioactive peptides whose structure was fairly easy to determine. The prosthetic group turned out to be remarkably similar to coenzyme A: the acyl groups were bound covalently to the thiol group of 4-phosphopantetheine, which in turn was bound through its phosphate group to a serine hydroxyl of the protein (Fig. 2.3). To study the biosynthesis of ACP, Vagelos made use of a strain of *E.coli* that had to be supplied with pantothenate in order to grow. When the bacteria were fed with radioactive pantothenate, the labelled substance was incorporated into the cell's ACP and CoA. However, whereas the ACP level remained constant under all conditions the level of CoA was very much dependent on the concentration of pantothenate. When the cells were grown in a medium of high pantothenate concentration, the level of CoA produced was also high; but if these cells were washed free of pantothenate and transferred to a medium with no pantothenate, then the CoA level dropped rapidly while that of ACP remained constant. In other words, ACP is synthesized at the expense of CoA. It seems that ACP synthesis is under tight control and this may be yet another factor involved in the overall control of fatty acid synthesis.

After the discovery and characterization of ACP, the different steps in the reaction sequence were quickly elucidated and the enzymes isolated and purified as illustrated in Table 2.6.

An important feature of the *E.coli* synthetase is that the products are mainly unsaturated fatty acids (vaccenic acid, $\Delta^{11}C_{18:1}$ accounts for about 70% of the total acids, the rest being saturated, mainly palmitic), in contrast to the yeast and animal synthetases which produce only saturated acids. This is because at the *dehydrase* step (step 5 in Table 2.6) an alternative reaction may occur: the *trans*-2 double bond may be isomerized to a *cis*-3 bond which is *not* reduced to a saturated acid by the reductase, but *is* elongated by further condensation with malonyl-ACP. We shall discuss this reaction under 'unsaturated fatty acid biosynthesis'. It is not clear what the final product of the *E.coli* synthetase is. The possibilities are: (*i*) acyl-ACP, which may be used directly for complex lipid synthesis (chapter 4); (*ii*) acyl-CoA — by transfer from ACP, or (*iii*) non-esterified fatty acid—by hydrolysis of acyl-ACP or acyl-CoA. Nor is it known what determines the chain length of the final product.

The animal and yeast synthetases are tightly organized complexes which cannot be fractionated into active constituent enzymes.

The binding of the enzymes of the animal and yeast synthetases in a tight complex has presented difficulties in studying the intermediate steps in these organisms. It does, however, have many advantages for the organism in that the intermediates can be passed on more efficiently from one enzyme to another rather like a conveyor belt. This is a feature of the organization of many enzyme systems in the more highly evolved organisms. (However, at least one species of bacteria has recently been discovered with a 'tightly bound' fatty acid synthetase). The pigeon liver synthetase has a molecular weight of 540,000 and contains 50 SH-groups per mole of complex. The American biochemist J.W. Porter has demonstrated that 4-phosphopantetheine is bound to one of the constituent proteins of the complex, indicating that the highly organized synthetases have an ACP-like component. The complex is unstable in solutions of low ionic strength, high pH or in presence of urea or deoxycholate and breaks down into small inactive 'sub-units'.

The steps catalysed by the yeast synthetase were demonstrated by the use of 'model substrates'.

The steps catalysed by the yeast synthetase were worked out by the German biochemist, F. Lynen, even though the intermediates could not be isolated. It was partly for this work that he was awarded the Nobel Prize in 1964. In *E.coli*, the individual intermediates could be isolated, purified and characterized. Then, in order to demonstrate any particular enzymic step, the purified intermediate (either isolated from the previous step, or chemically synthesized), could be used as a substrate for that step. In yeast,

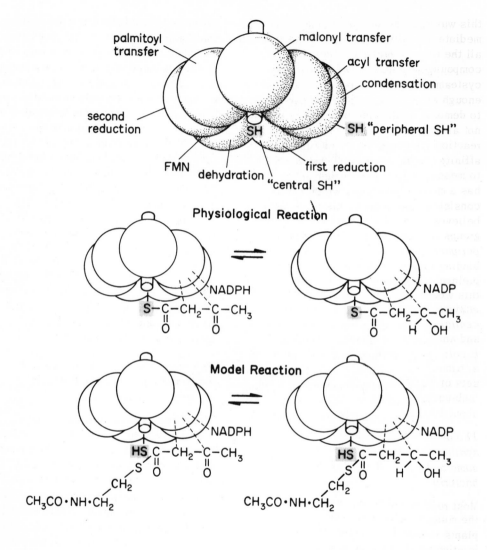

Fig. 2.5. Yeast fatty acid synthetase.

this was not possible because the intermediates remained bound to the enzymes all the time. Lynen therefore used 'model compounds' – acyl derivatives of N–acetyl cysteamine and hoped that there would be enough affinity for the enzyme to be able to demonstrate the reaction. This was not always the case, but in certain reactions the model substrates had enough affinity (though very small) to be able to measure the reaction. The yeast complex has a molecular weight of $2 \cdot 3 \times 10^6$ and consists of 7 enzymes (Fig. 2.5). Lynen believes that there are two types of SH–groups involved in the complex: a *peripheral* one which is sensitive to SH-binding reagents and is 'protected' by preincubating the enzyme with acetyl-CoA; this would be analogous to the *condensing enzyme* of the *E.coli* synthetase; also a *central* one, carried on 4-phosphopantetheine and analogous to the ACP component of *E.coli*. The active synthetase occurs as a 'trimer'; in other words, there are three sets of 7 enzymes in the complex or 21 'sub-units' in all, each having a M.W. of about 100,000.

The overall chemistry of fatty acid synthesis in higher plants is the same as in animals, yeasts and bacteria.

Most research so far has indicated that the major site of fatty acid synthesis in plants is the chloroplast. The process is stimulated in the light because photosynthesis generates the ATP required for acyl activation and the NADPH needed in the reduction steps. Acetyl-CoA carboxylase has been demonstrated in plant tissues but an alternative route to malonyl-CoA is also available by the

activation of malonate derived from oxaloacetate:

$$\text{Oxaloacetate} \xrightarrow{\text{CO}_2} \text{malonate} \longrightarrow \text{malonyl-CoA}$$

It is not clear yet whether in the physiological state the plant synthetases are multienzyme complexes as in yeast or dispersed like that of *E.coli*, but a protein very similar to bacterial ACP has been isolated from the soluble fatty acid synthesizing systems of lettuce and avocado pear by P.K. Stumpf's research team in California. They found that the products of a given *synthetase* may be altered by interchanging plant and bacterial ACP's. However, we are still a long way from knowing how the structure of the protein influences the role of ACP in fatty acid synthesis.

Branched chain fatty acids occur in many bacteria and originate either from branched chain amino acids or are formed by introduction of methyl groups into unsaturated fatty acids.

Most of the fatty acids with branches in the chain occur in bacteria. *Micrococcus lysodeikticus* for example is rich in C_{15} acids of both the *iso* type, 13-methyl-C_{14} or the *anteiso type*, 12-methyl-C_{14}. These have been shown to originate from the amino acids leucine and *iso*-leucine respectively by the American, W.J. Lennarz and the Canadian, T. Kaneda.

Another common branched chain fatty acid is 10-methyl stearic acid, *tuberculostearic* acid, a major component of the fatty acids of *Mycobacterium phlei*. In this case, the methyl group originates from the 'methyl donor' S-adenosyl

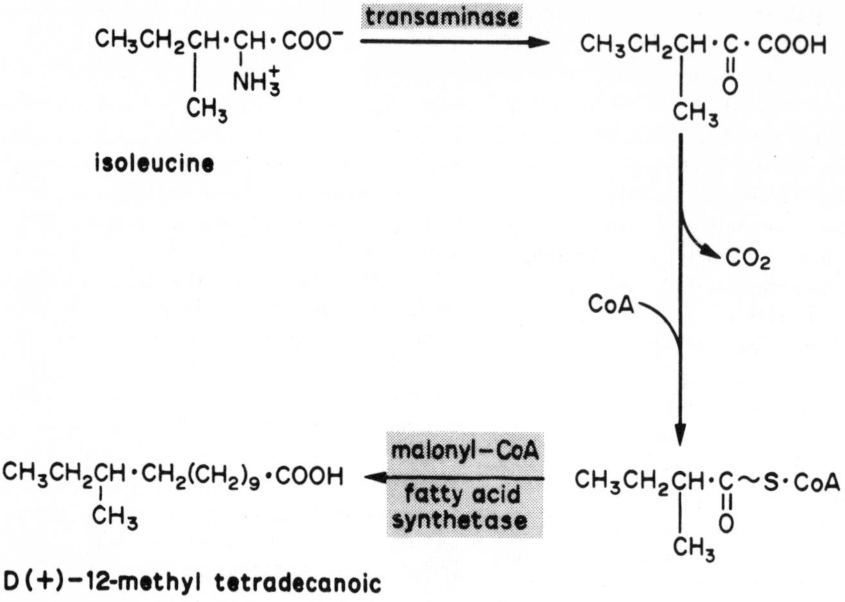

$$CH_3CH_2\overset{\underset{|}{CH_3}}{\underset{|}{\overset{|}{NH_3^+}}}CH \cdot CH \cdot COO^- \xrightarrow{\textbf{transaminase}} CH_3CH_2\overset{\underset{|}{CH_3}}{CH} \cdot \overset{\overset{O}{\|}}{C} \cdot COOH$$

isoleucine

$\searrow CO_2$

$CoA \searrow$

$$CH_3CH_2\overset{\underset{|}{CH_3}}{CH} \cdot CH_2(CH_2)_9 \cdot COOH \xleftarrow{\quad \boxed{\begin{array}{c}\textbf{malonyl}-\textbf{CoA}\\ \textbf{fatty acid}\\ \textbf{synthetase}\end{array}}\quad} CH_3CH_2\overset{\underset{|}{CH_3}}{CH} \cdot \overset{\overset{O}{\|}}{C} \sim S \cdot CoA$$

D(+)–12-methyl tetradecanoic

Fig. 2.6. Biosynthesis of branched chain fatty acids from amino acids.

methionine, while the 'acceptor' is oleic acid. The reaction takes place in two steps; the first product is 10-methylene stearic acid which is then reduced to 10-methyl stearic acid :

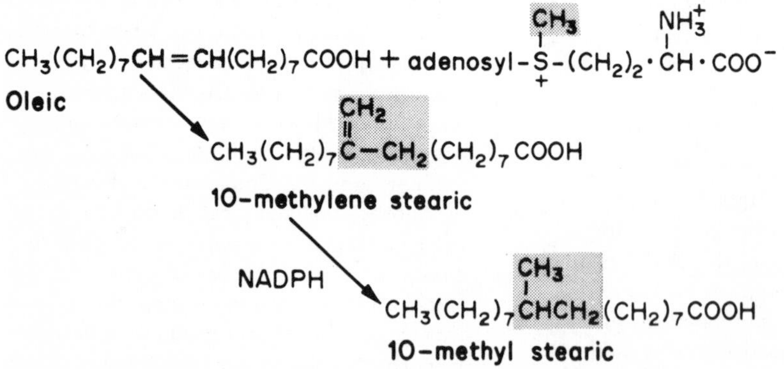

It is quite probable that this reaction takes place while the fatty acid is esterified in a phospholipid rather than in thiol ester linkage with CoA or ACP. The *Mycobacteria* also contain a range of

$$CH_3(CH_2)_7CH=CH(CH_2)_7COOH + \text{adenosyl}-\overset{+}{\underset{}{S}}\text{-}(CH_2)_2 \cdot \overset{\underset{|}{\overset{|}{NH_3^+}}}{CH} \cdot COO^-$$

$$\overset{\overset{\displaystyle CH_3}{|}}{}$$

Oleic

$$\searrow$$

$$CH_3(CH_2)_7\overset{\overset{\displaystyle CH_2}{\|}}{C}-CH_2(CH_2)_7COOH$$

10-methylene stearic

NADPH $\searrow$

$$CH_3(CH_2)_7\overset{\underset{}{\overset{\displaystyle CH_3}{|}}}{CH}CH_2(CH_2)_7COOH$$

10-methyl stearic

Fig. 2.7. Formation of branched chain fatty acids by methylation of unsaturated acids.

highly branched acids whose bio- synthesis is different from those we have considered so far. Acids such as *mycocerosic* are formed by the so-called 'propionic acid mechanism' in which a normal long chain fatty acid undergoes repeated condensations with propionic acid:

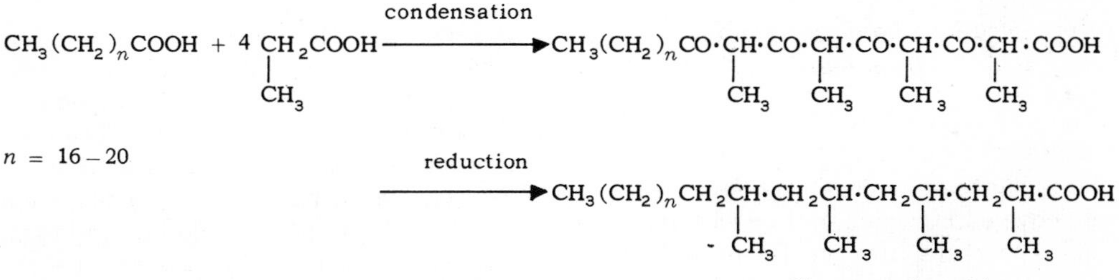

chain fatty acids, in some instances true reversal of β—oxidation does occur. The anaerobic bacterium *C. kluyveri* produces acetyl-CoA from ethanol, first by oxidation to acetaldehyde with *alcohol dehydrogenase*, then by a second oxidation to acetyl-CoA. This represents yet another alternative pathway to acetyl-CoA. The organism

$$CH_3(CH_2)_n COOH + 4\ CH_2COOH \xrightarrow{\text{condensation}} CH_3(CH_2)_n CO \cdot CH \cdot CO \cdot CH \cdot CO \cdot CH \cdot CO \cdot CH \cdot COOH$$

with CH$_3$ groups.

$n = 16-20$

$$\xrightarrow{\text{reduction}} CH_3(CH_2)_n CH_2 CH \cdot CH_2 CH \cdot CH_2 CH \cdot CH_2 CH \cdot COOH$$

Mycocerosic acid

A second type of branched chain acid, such as *corynomycolic* acid is formed by the 'head to tail' condensation of two molecules of palmitic acid :

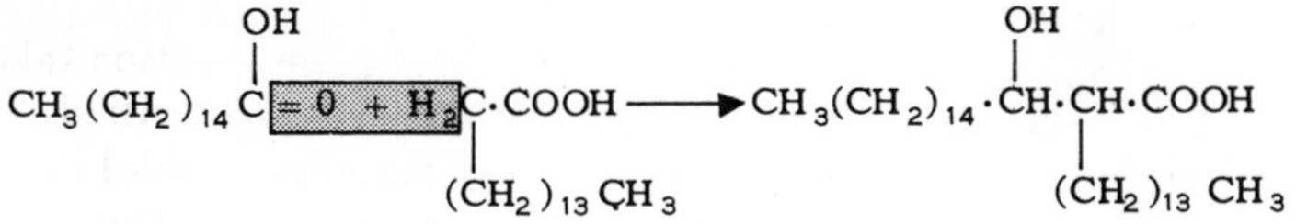

then synthesizes fatty acids up to $C_{6:0}$ by reversal of β-oxidation (see page 65). W. Seubert has also demonstrated β-oxidation reversal in soluble extracts

$$CH_3(CH_2)_{14} C{=}0 + H_2C \cdot COOH \longrightarrow CH_3(CH_2)_{14} \cdot CH \cdot CH \cdot COOH$$

Corynomycolic acid

Both of these pathways were worked out by the French chemist and biochemist E. Lederer, and his colleagues.

Some examples of reversal of β-oxidation have been found.

Although in most organisms the *malonyl-CoA pathway* is the main route to long

of animal mitochondria. The pathway does not require CO_2 or malonyl-CoA, but involves an NADPH-requiring enoyl-CoA reductase rather than the NADH-specific enzyme which occurs in β-oxidation. The main differences between β-oxidation and the malonyl-CoA pathway for fatty acid synthesis are listed in Table 2.7.

Table 2.7. Differences between β-oxidation and fatty acid synthesis

	β-Oxidation	Synthesis
Acyl carrier	CoA	ACP
Redox coenzyme	NAD/NADH	NADPH/NADP
'Two-carbon unit'	Acetate	Malonate
Configuration of β-hydroxy intermediate	$L(+)$	$D(-)$
Location	Mitochondria	Soluble fraction

Preformed long chain fatty acids may also be elongated.

The pathways we have discussed so far represent *de novo* synthesis from C_2 units. Several mechanisms have been discovered by which preformed fatty acids (perhaps from the diet) may be modified to suit the needs of the individual tissue or organism. In one such system, discovered by S. J. Wakil in rat liver mitochondria, an existing long chain fatty acid is elongated by addition of acetyl-CoA units with either NADPH or NADH as reducing coenzyme. Neither CO_2 nor malonyl-CoA is involved in this pathway. Alternatively, the Dutch biochemist D. Nugteren discovered that in a rat liver *microsomal* fraction, long chain fatty acids can be elongated by *malonyl-CoA* in presence of NADPH. In each of these mechanisms, unsaturated as well as saturated fatty acids can be elongated.

The details of the control of fatty acid synthesis are not yet known.

During our discussion of the various steps of fatty acid biosynthesis, we have briefly mentioned possible ways in which control may be exerted. By control we mean ways in which the rate or extent of synthesis

is regulated. How does the organism know when it needs to produce more or whether it has enough? What tells the organism to stop at a certain chain length? The places in a pathway where control is usually exerted are at the beginning or at the end. The first step in fatty acid biosynthesis, catalysed by acetyl-CoA carboxylase, is the rate limiting step. Both the *carboxylase* and the *synthetase* are inhibited by the end-product, for example palmitoyl-CoA, and this has been suggested as a possible control mechanism ('feedback inhibition'). However, acyl-CoA derivatives inhibit a great many enzymes unrelated to fatty acid synthesis, probably because they are powerful detergents. Citrate activates *acetyl-CoA carboxylase*, while *citrate synthetase* appears to be inhibited fairly specifically by acyl-CoA (acyl-dephospho-CoA does not inhibit the enzyme). These two intermediates, acyl-CoA and citrate, may therefore have some physiological significance in the control of fatty acid synthesis. We do not know whether citrate activates the enzyme *in vivo*. The reason for doubt is that a rather high concentration is needed for activation *in vitro*. As most of the cell's citrate is in the mitochondria while the enzyme is outside, the amount of control which could be exerted would seem to be limited.

When an animal is starved (or in the disease of diabetes), fatty acid synthesis is inhibited: the system is 'switched on' again by refeeding (or treatment with insulin). This 'switching on process can be prevented by administering substances which inhibit protein synthesis. This suggests that refeeding stimulates the production of new enzyme protein or we say that the enzymes may be *induced*. Another point of regulation may be in the activity of ACP hydrolase which could control the amount of ACP available. The relative importance of enzyme induction enzyme 'turnover', level of enzyme activity and 'feedback inhibition' is a subject awaiting more research.

The biosynthesis of hydroxy fatty acids

Hydroxy fatty acids may be formed as by-products or end products of the various pathways of fatty acid oxidation. Thus, *α-oxidation* can yield α- or 2-hydroxy acids which are major constituents of some lipids, particularly the brain glycolipids, (Chapter 5). The enzymes of *ω-oxidation* and related *hydroxylases* can yield fatty acids with hydroxyl groups at the methyl end of the chain. In these reactions the hydroxyl group is introduced into a pre-formed long chain fatty acid and the various oxidation pathways will be discussed more fully in the later section on fatty acid oxidation (pages 74, 76). Certain β- or 3—hydroxy acids, which are major constituents of some bacterial cell walls, probably arise as by-products of the normal pathway of fatty acid synthesis in bacteria.

One of the most important naturally occurring hydroxy acids is *ricinoleic* acid (*D*-12-hydroxy oleic acid, see Table 2.5) which accounts for 90% of the triglyceride fatty acids of castor oil and about 40% of the glyceride fatty acids of ergot oil, the lipid produced by the parasitic fungus *Claviceps purpurea*. In developing castor seed, ricinoleic acid is synthesized by hydroxylation of oleoyl-CoA in the presence of molecular oxygen, NADH and iron. Although the enzyme has many of the properties of a *mixed function oxygenase* (see p. 77), the mechanism contrasts with ω-oxidation in requiring the activated form (CoA thiol ester), but not requiring cytochrome P_{450} as a cofactor. In contrast to the hydroxylation mechanism in castor seed, the pathway in *Claviceps* involves *hydration* of linoleic acid under anaerobic conditions. Thus the hydroxyl group in this case originates from water and not from molecular oxygen. A similar hydration mechanism in the bacterium *Pseudomonas* produces, stereospecifically, *D*-10-hydroxystearic acid. This hydration appears to be reversible.

The biosynthesis of unsaturated fatty acids

Monoenoic acids are produced by two completely different pathways. The first, which occurs in anaerobic, and some aerobic bacteria involves an anaerobic mechanism.

Basically, there are two completely different pathways by which unsaturated fatty acids are produced. In an earlier section, we mentioned that the fatty acid synthetase of *E.coli*, in contrast to the mammalian and yeast synthetases, produced unsaturated as well as saturated acids. An idea of how this might arise was first put forward by the American biochemist, K. Bloch, who was studying the

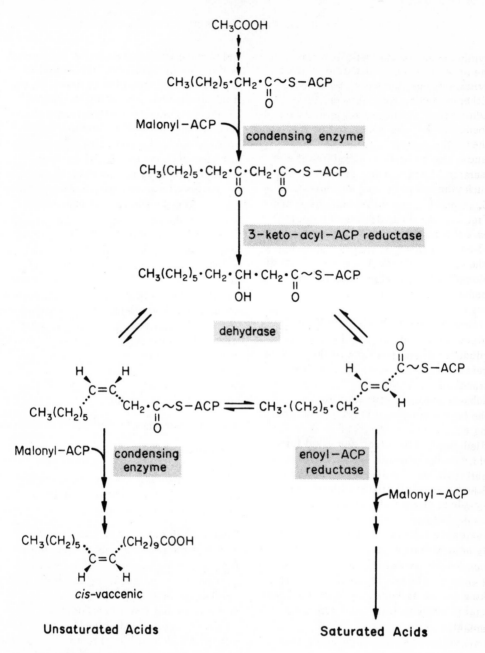

Fig. 2.8. The anaerobic pathway.

biosynthesis of unsaturated fatty acids in the anaerobic bacterium *Clostridium butyricum*. This organism, like almost all other bacteria produces only *mono-unsaturated* fatty acids. Two pairs of monoene isomers are produced: $\Delta7$ and $\Delta9$ hexadecenoic and $\Delta9$ and $\Delta11$ octadecenoic acids. The occurrence of these isomers can be explained if there are 'branch points' in fatty acid synthesis at $C_{8:0}$ and $C_{10:0}$ according to the scheme in Fig. 2.8. At these points, the normal *trans*-2-enoyl-ACP is isomerized to *cis*-3-enoyl-ACP which is not a substrate for the enoyl-ACP reductase, but *is* capable of elongation. Thus the *trans*-isomer formed at the branch point gives rise to a long chain *saturated* acid while the *cis*-isomer yields a long chain *cis-unsaturated* fatty acid. The positions of the double bonds in the final products, therefore, depend only on the positions of branch points.

Subsequent studies on the mechanism of the branching process have been done using *E.coli*, one of the most extensively studied organisms of the biochemical world. The fatty acid synthetase contains an activity which catalyses the dehydration of 3-hydroxy-decanoyl-ACP (β-hydroxy-decanoyl) to a mixture of *trans*-2 and *cis*-3-decenoates, and also the interconversion of the two double bond isomers. This enzyme is called *β-hydroxydecanoyl thioester dehydrase* and can be purified and separated from the synthetase by making use of the fact that the *dehydrase* is stable to heating at 50° for 10 min, whereas after 2 min of this treatment the *synthetase* is completely denatured. The puzzling fact emerged that the puri-fied enzyme produced mainly *trans*-2-enoyl-ACP (85%) whereas the complete synthetase produces predominantly unsaturated fatty acids — the elongation products of the less abundant *cis*-3-isomer. The most likely explanation of this apparent paradox is that the rate of C_2 addition to the *cis*-isomer is very much faster than the corresponding reductive step of the *trans*-2-isomer, thereby resulting in an accumulation of the unsaturated products.

In most organisms, unsaturated fatty acids are formed by an oxidative mechanism.

The *anaerobic* mechanism for mono-unsaturated fatty acid synthesis is by no means the most important. Indeed it has only been observed in a limited number of bacteria — the *Eubacteriales*. By far the most *widespread* pathway is by an oxidative mechanism, also discovered by Bloch's team, in which a double bond is introduced *directly* into the preformed saturated long chain fatty fatty acid with oxygen and NADPH as cofactors. This pathway is almost universal and is used by yeasts, higher animals, protozoa, algae and bacteria. Apparently the two pathways are mutually exclusive because no organism has been discovered in which they exist side by side. The oxidative pathway, too, exhibits remarkable specificity, as most of the acids produced have a $\Delta9$ double bond. Exceptions are the $\Delta7$ bonds in some algae, $\Delta5$ in *Bacillus megaterium*, a $\Delta10$ acid in *Mycobacterium phlei*, and a $\Delta6$ acid (petroselenic) in some plants.

The pathway was first demonstrated

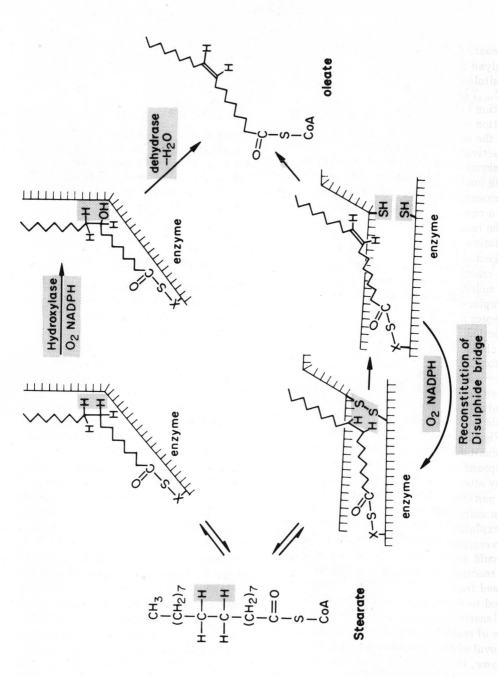

Fig. 2.9. Hypothetical schemes for oxidative desaturation.

in yeast. Cell-free preparations could catalyse the conversion of palmitic into palmitoleic acid (hexadec-9-enoic acid, $\Delta^9C_{16:1}$) only if both a particulate fraction (microsomes) and the supernatant fraction were present. Later it appeared that the supernatant contained the enzyme for activating the fatty acid to its coenzyme A derivative. The particles alone could perform the dehydrogenation. An examination of the cofactors required and a comparison of the characteristics of the reaction with other, better known oxidative reactions of a similar type suggested to Bloch a possible mechanism. The reaction required NADH or NADPH and molecular oxygen; the gas could not be replaced by an artificial electron acceptor as is the case for many oxidation-reduction reactions. These requirements are characteristic of a wide variety of hydroxylation reactions of the type R.H $\longrightarrow$ R.OH, catalysed by enzymes known as *oxygenases*. In the case of desaturation, the analogous pathway would be that shown in the upper part of Fig. 2.9. This mechanism would necessitate the formation of an hydroxy compound as an intermediate, but so far many attempts by biochemists to prove the participation of such compounds have been unfruitful. Such a failure might be explained in terms of an hydroxy acid *irreversibly* bound to the enzyme so that it could not therefore be isolated during the reaction, nor, conversely could it be formed from exogenous hydroxy acid added to the system. An alternative explanation is for a completely different type of reaction involving the simultaneous removal of two hydrogen atoms by the enzyme, the molecular oxygen being

needed at a later stage to re-oxidize some as yet unknown intermediate which has been reduced in the process. A *highly speculative* scheme proposed in the Authors' laboratory is shown in the lower half of Fig. 2.9. The precise chemical mechanism of this process is one of the most fascinating problems to be unravelled in the field of unsaturated fatty acid biochemistry today. In spite of our lack of knowledge of the mechanism, certain details have emerged in recent years, mainly concerning the stereochemistry of the reaction. The removal of hydrogen atoms in bacterial, animal and plant desaturations has been shown to be absolutely stereospecific in that the *D*-9 and *D*-10 (*cis*) hydrogen atoms are removed during the desaturation.

Plants were once thought to possess a unique pathway.

It is somewhat unusual for the constituents of cells to arise by multiple pathways. Ordinarily they arise by a single pathway whose constituent reactions do not differ very much from organism to organism. It was all the more surprising, therefore, to come across yet a third pathway for mono-unsaturated fatty acid synthesis in higher plants. However, the evidence for this third pathway is tenuous and has been put forward, not so much because there is firm positive evidence for its existence, but more because people have failed to demonstrate the other two pathways. While higher plants could synthesize long chain monounsaturated fatty acids if the precursor were acetate, they apparently could not directly desaturate the corresponding saturated acids or their CoA

ANIMALS

De Novo Synthesis

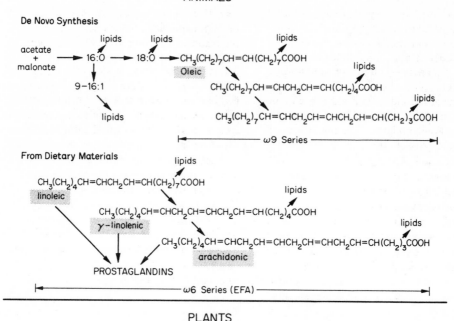

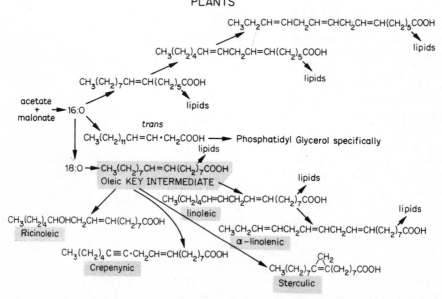

Fig. 2.10. Some important pathways of fatty acid biosynthesis in animals and plants.

thiol esters. Part of the mystery surrounding the formation of oleic acid in plant leaf chloroplasts (the major site of synthesis of leaf fatty acids) was removed by two sorts of experiments. In the first (performed by James's group) leaf tissue was allowed to synthesize fatty acids, starting from [^{14}C]-acetate under anaerobic conditions, when only palmitic and stearic acids were formed. When such labelled tissue was transferred to air, part of the labelled stearic acid disappeared and was replaced by a roughly equivalent amount of labelled oleic acid. In the second type of experiment, Bloch showed that leaf chloroplast preparations were able to convert stearoyl-S-ACP into oleic acid. Indeed, the phytoflagellate, *Euglena gracilis*, which can behave as an animal or as a plant according to its environment, has dual substrate specificities — it can desaturate only stearoyl-S-CoA if grown heterotrophically or stearoyl-S-ACP when grown photosynthetically. It appears, therefore, that there may not be any mechanistic distinction between the 'plant' pathway and the oxidative mechanism, except that either the plant contains distinct pools of stearic acid not able to mix with added stearate or lacks the enzyme required to transfer stearate from the CoA form (which we know can incorporate its acyl group into the lipids) to ACP, which we infer may be needed for desaturations.

The cyclopropene fatty acid, sterculic acid, when supplied in the diet of laying fowls, gives rise to a condition known as 'pink-white disease'. The only defined action of this acid is the inhibition of the conversion of stearic acid into oleic acid and may be connected with the known reaction of the cyclopropene ring with —SH compounds. In plants, however, sterculic acid (see Table 2.5) occurs together with those unsaturated acids known to be formed by sequential desaturation of stearic acid and it therefore has no inhibitory action. This evidence also points to a difference in the ways in which plants and animals handle stearic acid so far as desaturation is concerned.

An unusual monounsaturated acid, specifically esterified in phosphatidyl glycerol of photosynthetic tissue, is *trans*-3-hexadecenoic acid. This is formed from palmitic acid by direct oxidative desaturation but the mechanism has not yet been worked out, (see Fig. 2.10).

Animals and plants make different families of polyunsaturated fatty acids, and bacteria, none.

Except for bacteria, all organisms are capable of introducing more than one double bond into the chain. These double bonds are always separated by a methylene group and are known as '*methylene-interrupted*' polyunsaturated fatty acids. Some bacteria *do* have polyunsaturated fatty acids but not of the methylene interrupted type, and the biosynthetic pathways leading to these acids have not been worked out yet. In animals, the second and subsequent double bonds are introduced between the first double bond and the carboxyl group, whereas in in plants the new double bonds are introduced toward the methyl end. Some primitive organisms intermediate in the evolutionary scale, such as the phytoflagellate *Euglena*, have the ability to desaturate in either direction. The

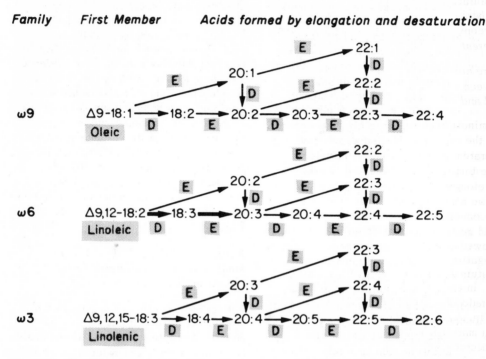

Fig. 2.11. The three important families of polyunsaturated fatty acids in animals showing chain elongation (E) and desaturation (D).

result of this is that distinct families of polyunsaturated fatty acids arise, which can be more easily recognized by numbering the double bonds from the methyl end, than the carboxyl end of the chain. Hence the three most important families of polyunsaturated fatty acids are $\omega 9$, $\omega 6$ and $\omega 3$, and the first members of each family are oleic, linoleic and α-linolenic respectively (Table 2.4 and Fig. 2.11).

Linoleic acid ($\Delta 9, 12-18:2$), the first member of the $\omega 6$ series, and certain $\omega 6$ acids derived from it, are necessary to maintain animals in a healthy condition; the $\omega 9$ isomer ($\Delta 6,9-18:2$) is inadequate in this respect. The inability of animals to desaturate oleic acid toward the methyl end of the chain, therefore, means that linoleic acid must be supplied in the diet from plant sources. Linoleic acid and related $\omega 6,9$-acids are known as *essential fatty acids* and will be discussed in the next section. As can be seen in

Fig. 2.11, animals can convert dietary linoleic acid by a series of alternate desaturations and elongations into *other members of the* $\omega 6$ series. However, interconversions between members of *different families* do not occur in animals.

There are alternative routes between one polyunsaturated acid and another.

Examination of Fig. 2.11 will show you that the conversion of one polyunsaturated fatty acid into another of the same family by alternate desaturations and elongations can sometimes proceed by two alternative pathways. For example, the conversion of $18:2\,\omega 6$ into $20:3\,\omega 6$ could go first by desaturation to $18:3\,\omega 6$ followed by elongation, or by an initial elongation to $20:2\,\omega 6$ followed by desaturation. Present evidence suggests that, in rats at least, the former pathway is predominant (indicated by heavy arrows in Fig. 2.11).

In mammals the predominant polyunsaturated fatty acid is usually arachidonic ($\Delta 5,8,11,14{-}20:4$; $20:4\,\omega 6$). Some higher unsaturated acids are formed by mammals, but the chief sources of these are fish, which have large amounts of 22:5 and 22:6 (Table 2.3). A summary of the pathways of fatty acid biosynthesis in plants and animals is given in Fig. 2.10.

The formation of the second and subsequent double bonds depends, like the first, on molecular oxygen and NADPH.

Like the oxidative desaturations leading to monounsaturated fatty acids, these further desaturations are firmly particle-bound, require oxygen and NADPH as cofactors, and involve the stereospecific removal of the D-hydrogen atoms in the formation of each double bond. The pathways in animals have been worked out by W. Stoffel in Germany, D.H. Nugteren in The Netherlands and the research groups of J.F. Mead and R.T. Holman in the U.S.A., and the enzymes are located in the membranes of the endoplasmic reticulum. Work in the Authors' laboratory has shown that in plants the reactions take place mainly in the chloroplast.

Competition between polyunsaturated fatty acids of different series for desaturases, elongating enzymes, and transacylases may control their biosynthesis.

Fatty acids rarely exist in the cell in the free form. The enzymes which incorporate them into complex lipids (see Chapter 4) are very active and, therefore, newly synthesized fatty acids, or those of dietary origin are rapidly esterified.

Before this can occur, they must be activated to their CoA or ACP derivatives, but before being incorporated into specific lipids, they may be modified by desaturation, chain elongation, or both. A fatty acid has a 'choice', therefore, of a number of competing reactions and competition between the enzymes of these various pathways serves to exert a control over the final fatty acid spectrum in any one kind of cell. Members of the different families may inhibit the desaturations in another family, *in vitro*, a fact which has been interpreted as due to competition for the same site on a desaturase. Conversely, the presence of some acids, particularly $20:4\omega6$, increases the rate of desaturation of the other members of a series, possibly by monopolizing one further source of competition — namely the acylating enzymes (see Chapter 4). It has been suggested from these data that there may be two types of desaturases — one for introducing the first double bond and another single enzyme for further 'polydesaturations'.

Different types of polyunsaturated acids accumulate on different classes of lipids.

The different types of polyunsaturated fatty acids accumulate, too, in different classes of complex lipids. For example, an organism like *Euglena*, which can live as a plant or an animal and which can synthesize both types of polyunsaturated fatty acids, accumulates its animal-type fatty acids (e.g. arachidonic) in the phospholipids, and its plant-type acids (e.g. α–linolenic) in the galactolipids. Observations of this sort have led to an hypothesis that certain complex lipids may play a direct role in the mechanism of desaturation and a specific example, in which oleoyl-phosphatidyl choline (see chapter 4) is converted directly into linoleoyl-phosphatidyl choline in *Chlorella* chloroplasts was recently demonstrated by the Authors.

The nature of the desaturase enzymes.

One of the facts that any theory about the mechanism of desaturation has to explain, is the quite remarkable *positional* specificity of the double bond. One can imagine that the fatty acid carboxyl group may form some kind of ester linkage (probably through a sulphur atom) on the enzyme, which would therefore 'fix' the carboxyl group in a certain position. Assuming a fairly restricted confirmation of the hydrocarbon chain, then an *active site* at a position of, say, 9–10 carbon atoms from the *binding site* will always ensure a $\Delta9$ double bond position. If this model is correct, two possibilities can arise: either that there is a single enzyme with *active sites* at the $\Delta6$, and $\Delta9$ and $\Delta12$ positions or several enzymes, each having a single site at a specific position. And again, how is it that some organisms have the enzymes for creating a $\Delta9$ *and* a $\Delta7$ monoene? Has the latter enzyme *evolved* from the former by elimination of one amino acid from the chain (a deletion which might approximately account for a shift in double bond position of two carbons)?

Little or nothing is known about the mechanism by which two hydrogen atoms are removed to form a double bond. The molecular oxygen and pyridine nucleotide requirements have still to be explained. Two possible mechanisms were discussed earlier, but it is unlikely that many more details of the mechanism will be discovered until the enzyme can be obtained in a purer state, free from the many other contaminating activities which confuse the picture. The nearest approach to the problem has been made by Bloch who had fractionated the 'soluble' desaturase from the photosynthetic micro-organism *Euglena gracilis* into three components, the *desaturase* itself, an enzyme called *NADPH-oxidase* and an iron and sulphur containing protein of very low redox potential, known as ferredoxin. Desaturating enzymes of all other tissues such as liver are tightly bound as part of the network of membrane structures known as the *endoplasmic reticulum* — a fact which makes them difficult to isolate and purify. Quite recently, the authors described a way in which these membranes may be broken down to smaller particles, so that in future it may be possible for the animal desaturase to be 'taken to pieces' as well. In any case, the elucidation of this mechanism provides the greatest challenge in the field of unsaturated fatty acid biosynthesis today.

Essential fatty acids and prostaglandins

The great period of isolation and identification of vitamins lay between 1840 and the late 1920's. During this time the deficiency effects of all the water-soluble and lipid-soluble compounds were demonstrated, though, not surprisingly, elucidation of their structures and total synthesis took another 35yr. By the end of the 1920's it was thought that all the major accessory food factors had been discovered and that carbohydrate and fat were important only in so far as their calorie contribution was concerned.

This whole concept was overturned in 1930 when the Americans G.O. Burr and M.M. Burr described how acute deficiency states could be produced in rats by feeding fat-free diets and that these deficiencies could be eliminated by adding

only certain specific fatty acids to the diet. It was shown that linoleic and later arachidonic acids were responsible for this effect and the term *Vitamin F* was coined for them. However, this name has now been reserved for some factors of the vitamin B complex, so that the term *essential fatty acid* (EFA) is now *de rigueur*.

Unsaturated fatty acids having a specific double-bond structure —'ω-6,9'— are found to be essential for maintaining animals in a healthy condition.

Essential fatty acid deficiency can be produced in a variety of animals, including man, but data for the rat (the major experimental animal) are the best documented and the effects are summarized in Table 2.8. The effects of different acids on relief of deficiency symptoms were first compared by measurement of growth response, but later H.J. Thomasson at the Unilever Laboratories in The Netherlands used a test based on the disturbances in water metabolism. On this basis, the relative effects of the natural di–, tri– and tetra–unsaturated acids showed that arachidonic and linoleic acids had the same order of activity whereas the activity of α–linolenic acid $(9,12,15-18:3)$ was much lower. γ–Linolenic acid $(6,9,12-18:3)$ on the other hand was as active as linoleic and an $11,14-20:2$ acid had moderate activity. Biochemists began investigating a wider variety of unsaturated fatty acids for their EFA activity. It is quite clear from their results that there is no common relationship between the position of the double

Table 2.8. The major effects of EFA deficiency in the rat

1. Skin symptoms	Dermatosis; increased water permeability; drop in sebum secretion; epithelial hyperplasia.
2. Weight	Decrease.
3. Circulation	Heart enlargement; decreased capillary resistance; and increased permeability.
4. Kidney	Enlargement; intertubular haemorrhage.
5. Lung	Cholesterol accumulation.
6. Endocrine glands	(a) *Adrenals.* Weight decreased in females and increased in males. (b) *Thyroid.* Reduced weight.
7. Reproduction	(a) *Females.* Irregular oestrus and impaired reproduction and lactation. (b) *Males.* Degeneration of seminiferous tubules.
8. Metabolism	(a) Changes in fatty acid composition of most organs. (b) Increase in cholesterol levels in liver, adrenal and skin. (c) Decrease in plasma cholesterol. (d) Changes in swelling of heart and liver mitochondria and uncoupling of oxidative phosphorylation. (e) Increased triglyceride synthesis and release by the liver.

bond measured from the carboxyl group and EFA activity (Table 2.9).

On the other hand, classification on the basis of double bond position measured from the methyl terminal group showed that all the active acids possessed double bonds in the ω-6,9-positions.

When the pathway for the biosynthesis of polyunsaturated fatty acids was worked out, the sequence:

ω–6,9 hypothesis through studies of a wide range of synthetic unnatural polyunsaturated acids (Table 2.9).

Despite so much elegant and painstaking work, no real clue emerged as to the function of these essential fatty acid. While no well documented disease state due to dietary deficiency in humans has been described, it has been shown experimentally with orphanage children in the U.S.A. by A. A. Hansen that skin

Linoleic → γ–linolenic → di–homo–γ–linolenic → arachidonic
ω–6,9–18:2 ω–6,9,12–18:3 ω–6,9,12–20:3 ω–6,9,12,15–20:4

fitted the 'ω–6,9' classification admirably. Further work by Thomasson's group in collaboration with the organic chemist D.A. van Dorp, extended and confirmed the

conditions similar to those produced in rats could be obtained in human beings on a fat-free diet. The skin lesions disappeared when linoleic acid was added

Table 2.9. Relation between structure and EFA activity

| Fatty acid chain length | Position of double bonds | | Potency units/g |
	From methyl end ($\omega-$)	From carboxyl end ($\Delta-$)	
C_{18}	6	12	0
C_{18}	9	9	0
C_{18}	6:9	9:12	100
C_{18}	9:12	6:9	9
C_{18}	6:9:12	6:9:12	115
C_{18}	6:9:12:15	3:6:9:12	34
C_{19}	6:9	10:13	9
C_{19}	6:9:12	7:10:13	6
C_{19}	6:9:12:15	4:7:10:13	20
C_{20}	6:9	11:14	42
C_{20}	6:9:12	8:11:14	102
C_{20}	6:9:12:15	5:8:11:14	139

to the diet. It is now accepted that an adult human being needs about 10g of linoleic acid per day. This is almost universally fulfilled, since in most countries there is sufficient material of plant origin in the diet.

The fact that animals need these specific fatty acids of plant origin, because they lack the enzyme to insert a $\Delta-12,13-$double bond has already been discussed in an earlier section. What is still obscure is why a $\Delta12,13-$double bond should be essential.

A new type of oxygenated unsaturated fatty acid – Prostaglandin – is discovered, minute amounts of which have the ability to stimulate smooth muscle.

While these biochemical and chemical studies were underway, a new dimension was added to the EFA story by work in a very different field. As early as 1930, two American gynaecologists, R. Kurzrok and C.C. Lieb, reported that the human uterus, on contact with fresh human semen, was provoked into either strong contraction or relaxation. Both U.S. van Euler in Sweden and M.W. Goldblatt in England subsequently discovered marked stimulation of smooth muscle by seminal plasma. Von Euler then showed that lipid extracts of ram vesicular glands contained the activity and this was associated with a fatty acid fraction. The active factor was named *Prostaglandin* and was shown to possess a variety of physiological and pharmacological properties.

In 1947, the Swede, S. Bergström, started to purify these extracts and soon showed that the active principle was associated with a fraction containing unsaturated hydroxy acids. The work then lapsed until 1956, when with the help of an improved test system (smooth muscle stimulation in the rabbit duodenum

Table 2.10. Distribution of the major prostaglandins in different tissues

Source		PGE_1	PGE_2	PGE_3	$PGF_{1\alpha}$	$PGF_{2\alpha}$	$PGF_{3\alpha}$
Vesicular gland:	sheep	+	+	+	+		
Seminal plasma:	sheep	+	+	+	+	+	
	human	+	+	+	+	+	
Menstrual fluid:	human		+			+	
Lungs:	sheep		+			+	
	bovine					+	+
	pig, guinea pig						
	monkey, human					+	
Iris:	sheep					+	
Brain:	bovine					+	
Thymus:	calf	+					
Pancreas:	bovine		+				
Kidney:	pig		+				

Bergström isolated two prostaglandins in crystalline form, (PGE_1 and $PGF_{1\alpha}$). Their structure, as well as that of a number of other prostaglandins was elucidated by a combination of degradative, mass spectrometric, X-ray crystallographic and N.M.R. studies (Fig. 2—12). The nomenclature is based on the fully saturated C_{20} acid with C_8 to C_{12} closed to form a 5-membered ring; this is called prostanoic acid. Thus PGE_1 is designated 9-keto-11α,15α-dihydroxyprost-13-enoic acid. The 13,14 double bond has a *trans* configuration; all the other double bonds are *cis*. Figure 2.12 clearly brings out the difference between the 'E' and 'F' series, which have a keto and hydroxyl group at position 9 respectively. 'α' refers to the stereochemistry of the hydroxyl.

Prostaglandins are formed by ring closure and oxygenation of essential polyunsaturated fatty acids such as arachidonic

After the structures of PGE_1 and $PGF_{1\alpha}$ had been defined, the subsequent rapid exploitation of this field, including the unravelling of the biosynthetic pathways was done almost entirely by two research teams: those led by D.A. van Dorp in Holland and by S. Bergström and B. Samuelsson in Sweden. Both realized that the most likely precursor of PGE_1 and $PGF_{1\alpha}$ was arachidonic acid. This was then demonstrated by both groups simultaneously (with the same sample of tritiated arachidonic acid) by incubation with whole homogenates of sheep vesicular glands. The free fatty acid is the effective precursor, not an 'activated' form. The ability of the arachidonic acid chain to fold allows the appropriate groups to come into juxtaposition for the ring closure to occur (Fig. 2.12, 2.13). The enzymes occur in the microsomal fraction but a heat stable factor from the supernatant is required. This cofactor can be replaced by reduced glutathione, tetrahydrofolate or 6,7-dimethyltetrahydropteridine, but not by NADH or NADPH. The reaction also requires

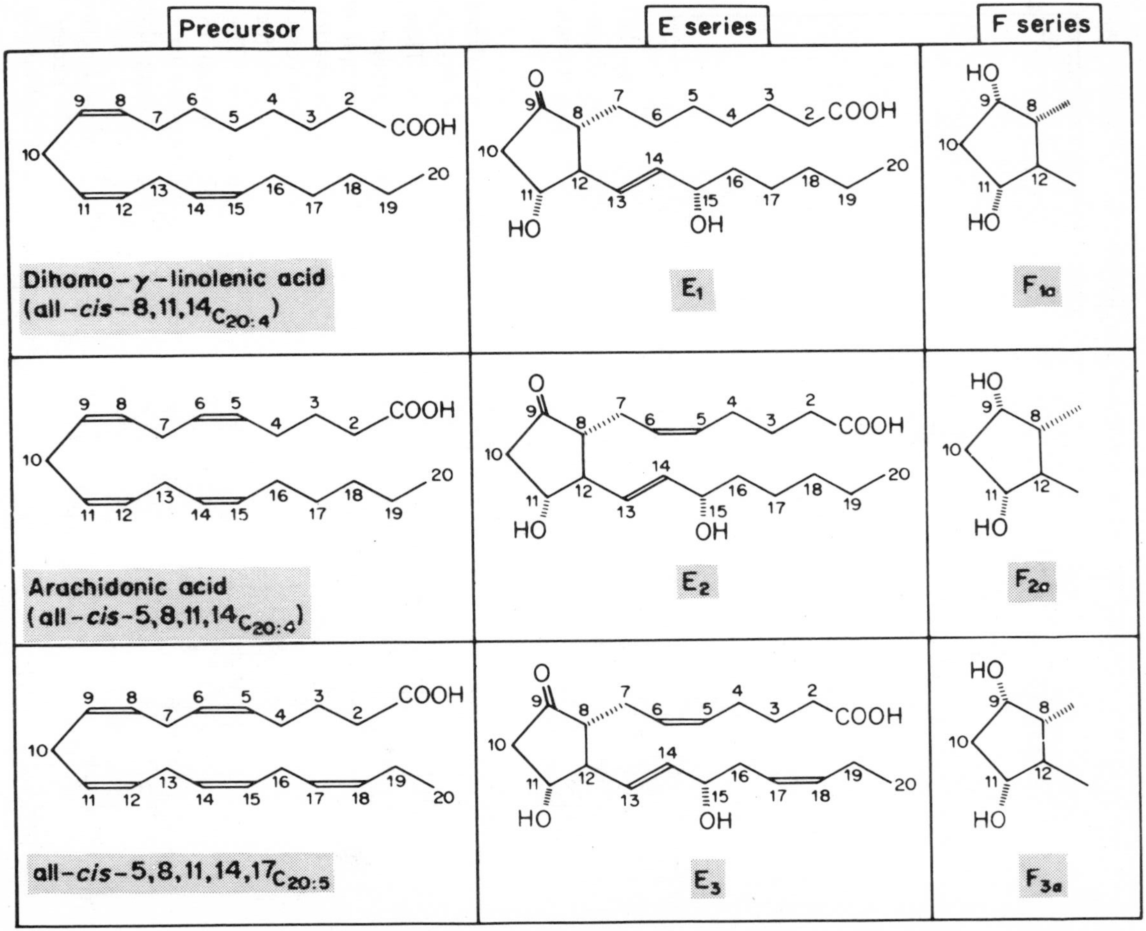

Fig. 2.12. Structure of prostaglandins E and F and their precursors.

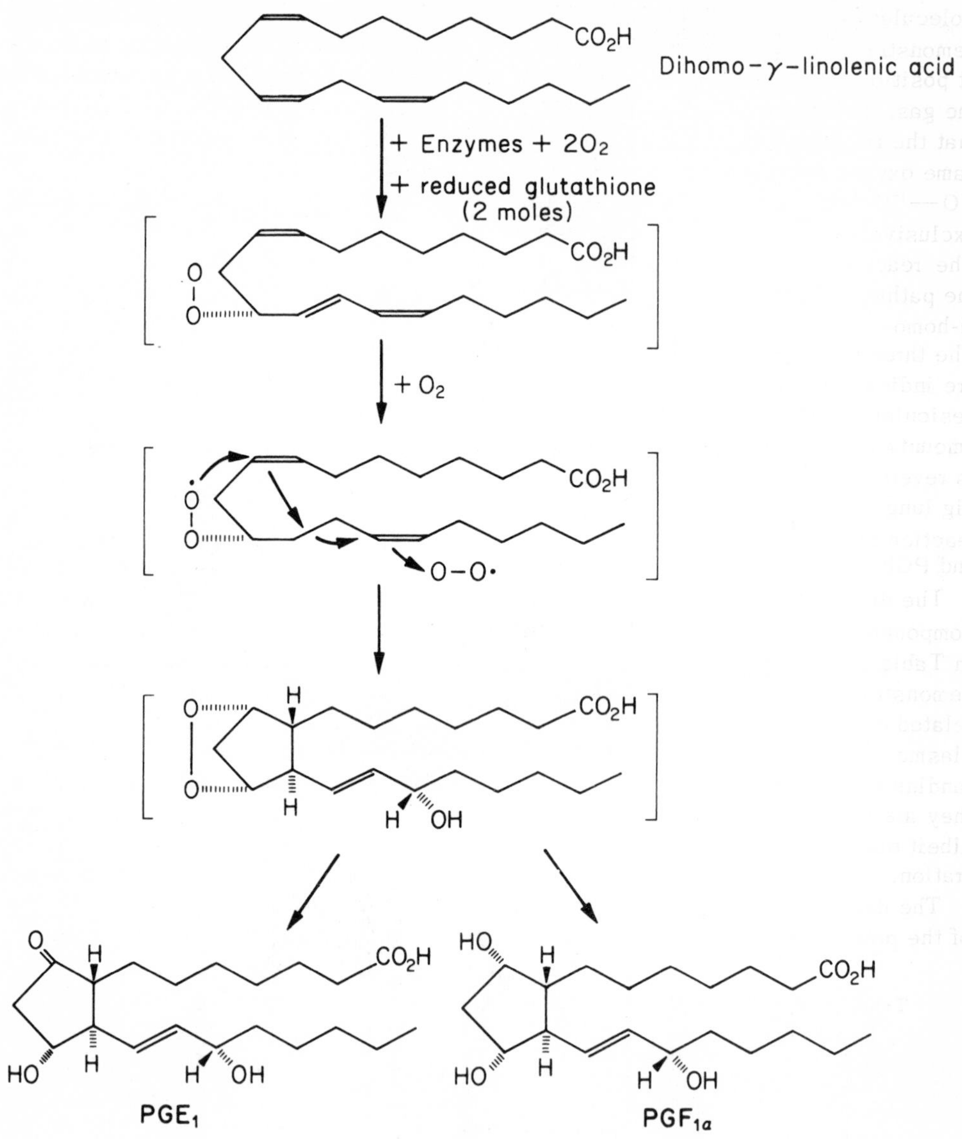

Fig. 2.13. Mechanism of biosynthesis of PGE$_1$ and PGF$_{1\alpha}$.

molecular oxygen. Labelling with $^{18}O_2$ demonstrated that all three oxygen atoms at positions 9,11 and 15 are derived from the gas. In addition, Samuelsson proved that the two ring oxygens came from the same oxygen molecule, since a mixture of $^{18}O—^{18}O$ and $^{16}O—^{16}O$ gave products exclusively either $9,11–^{18}O$ or $9,11–^{16}O$. The reaction is now believed to follow the pathway shown in Fig. 2.13 for di-homo–γ–linolenic acid as precursor. The three precursors of the PGE series are indicated in Fig. 2.12. Sheep vesicular glands give rise to only small amounts of the PGF compounds but this is reversed with homogenates of guinea pig lung. Fig. 2.13 shows the overall reaction in the system giving PGF_2 and PGE_2.

The distribution of the PGE and PGF compounds in a variety of tissues is shown in Table 2.10. More recent work has demonstrated an even greater range of related compounds in human seminal plasma. The richest source of prostaglandins is the seminal plasma, though they are found in almost all cell types, albeit sometimes at very low concentration.

The detailed pathways of breakdown of the prostaglandins have yet to be fully worked out, but lung homogenates can carry out both a reduction of the 13,14-double bond and a dehydrogenation of the 15-hydroxyl group. The relative amounts of the two reactions vary from animal to animal.

Prostaglandins do not have one but a variety of physiological effects. Only those fatty acids which give rise to biologically active prostaglandins have EFA activity but some of these do not have the 'ω–6,9' structure.

The prostaglandins exert a range of profound physiological activities at concentrations down to $10^{-9}g/g$ tissue. Apart from the early demonstrated effects on smooth muscle contraction, they lower blood pressure (by peritoneal vasodilation); decrease platelet 'stickiness', and in some cases (e.g. in the dog) increase blood pressure by peripheral venoconstriction. Continuous perfusion of very small amounts (e.g. $0\cdot01~\mu$g of PGE_1,/min) into pregnant women causes a similar uterine activity to that encountered in normal labour without any effect on the blood pressure. The nature of the effect varies from compound to compound (see Table 2.11). Apart from these actions,

Table 2.11. **Relative physiological activities of the prostaglandins**

Experimental system		PGE_1	PGE_2	PGE_3	Dihydro $-PGE_1$	PGF_1
Rabbit:	Duodenum	1	3·3	0·4	0·4	1·5
Guinea pig:	Ileum	1	—	0·2	0·1	0·05
Rabbit:	Blood pressure	1	0·6	0·3	0·6	0·1
Guinea pig:	Blood pressure	1	—	—	1·6	—
Rat:	Fat pad	1	0·1	0·2	0·1	0·1

PGE$_1$ is an effective antagonizer of the effect of a number of hormones on free fatty acid release from adipose tissues. This antagonism is believed to have as its site of action the cyclase enzyme that converts ADP to cyclic AMP. A particular interest lies in the effects of the prostaglandins on the cardio-vascular system, since it has been suspected for some time that coronary artery disease is connected in some way with a disturbance of lipid metabolism involving the essential fatty acids.

More recent work from the Dutch school and from R. Holman and his colleagues at the Hormel Institute in Minnesota has shown that only those fatty acids (including new synthetic odd numbered acids) that act as precursors to bio-logically active prostaglandins have EFA activity. However, the 'ω–6,9 hypothesis' has had to be dropped since some of the new synthetic acids do not contain this structure (Table 2.12). Although there is this correlation between EFA activity and the potential to be converted into a prostaglandin, one can-not cure EFA deficiency by infusion of any of the prostaglandins because of their rapid rate of destruction in the body. It still cannot be said with absolute certainty that the prostaglandins are responsible *inside* cells for all the biological activity of the essential fatty acids. Nevertheless, the indirect evidence is so strong that there can be little doubt that it is the prostaglandins that are the primary factors.

This field of work is rapidly expanding and in the next few years, prostaglandins or synthetic prostaglandin-like substances will probably be important therapeutic agents.

Table 2.12. Comparative EFA and prostaglandin activity of some new synthetic unsaturated fatty acids

Fatty acid		Position of double bonds		Activity	
Type ω	Chain length and number of double bonds	From carboxyl group	From methyl group	EFA	PG
4	18:3	8,11,14	4,7,10	−	−
5	19:3	8,11,14	5,8,11	+	+
6	20:3	8,11,14	6,9,12	+	+
7	21:3	8,11,14	7,10,13	+	+
8	22:3	8,11,14	8,11,14	−	−
4	18:4	5,8,11,14	4,7,10,13	−	−
5	19:4	5,8,11,14	5,8,11,14	+	+
6	20:4	5,8,11,14	6,9,12,15	+	+
7	21:4	5,8,11,14	6,10,13,16	+	+
8	22:4	5,8,11,14	8,11,14,17	−	−
5	20:3	9,12,15	5,8,11	−	−
7	20:3	7,10,13	7,10,13	−	−
8	20:3	6,9,12	8,11,14	−	−
9	20:3	5,8,11	6,9,12	−	−

Biohydrogenation of Unsaturated fatty acids

The introduction of double bonds into a fully saturated chain is a reaction widespread in nature. The reverse process, namely the hydrogenation of double bonds to produce more saturated fatty acids is rarer but occurs in the rumen of ruminant animals. This reaction is brought about by micro-organisms in the rumen. Linoleic acid, for example, is hydrogenated by rumen flora to stearic acid by the following series of reactions:

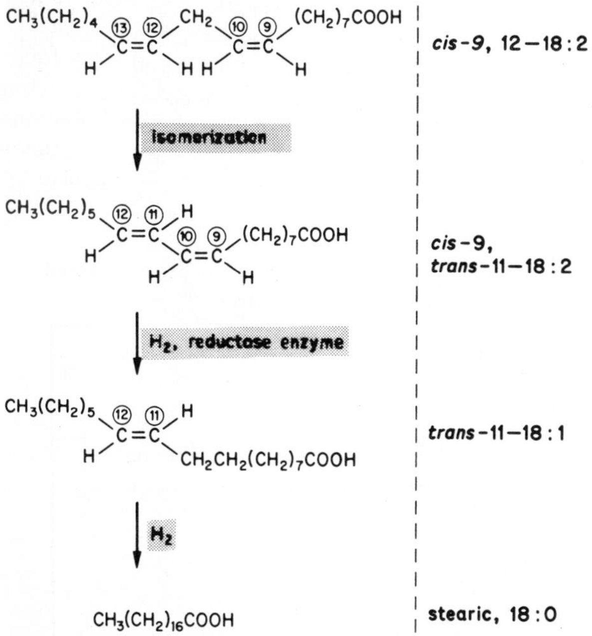

is more complicated than we have described here and a whole series of cis and trans positional isomers may be produced especially if the substrate is a more highly unsaturated acid such as linolenic or arachidonic acids.

The isomerization reaction can be demonstrated in cell-free extracts of the obligate anaerobe *Butyrivibrio fibrisolvens*, isolated from rumen flora, but so far, no cell-free system has been found to be capable of catalysing the hydrogenation step and the mechanism is unknown.

$CH_3(CH_2)_4$ (13) (12) CH_2 (10) (9) $(CH_2)_7COOH$
$C=C$ $C=C$
H H H H

cis-9, 12—18:2

↓ isomerization

$CH_3(CH_2)_5$ (12) (11) H
$C=C$ (10) (9) $(CH_2)_7COOH$
H $C=C$
H H

cis-9, trans-11—18:2

↓ H_2, reductase enzyme

$CH_3(CH_2)_5$ (12) (11) H
$C=C$
H $CH_2CH_2(CH_2)_7COOH$

trans-11—18:1

↓ H_2

$CH_3(CH_2)_{16}COOH$

stearic, 18:0

The first reaction involves an isomerization of the cis-12,13 double bond to form a trans-11,12 bond in conjugation with the cis-9,10 double bond. Next, hydrogen is added across the cis-9,10 bond to form trans-vaccenic acid and finally the trans-bond is also reduced to give the fully saturated fatty acid. In fact, the situation

The hydrogenation of unsaturated acids in the rumen occurs after they have been released from dietary lipids by hydrolytic enzymes known as *lipases*. Ruminants do not appear to suffer from EFA deficiency despite the fact that a large proportion of their dietary EFA are destroyed by biohydrogenation. The amount of unchanged EFA passing through the rumen (up to 4% of dietary intake) is sufficient for the needs of the animal.

The biosynthesis of cyclic acids

The only ring structures we shall discuss are the cyclopropanes and cyclopropenes. Little is known about the formation of the larger rings which occur in acids such as chaulmoogric.

The methylene group in cyclopropane acids originates from the methyl group of methionine in S-adenosyl methionine ('active methionine'). This is the same methyl donor which is involved in the formation of 10-methylene stearic and 10-methyl stearic acid from oleic acid (see p. 42). The 'acceptor' of the methyl group is likewise an unsaturated fatty acid. Thus, cis-vaccenic acid gives rise

to lactobacillic acid, while oleic acid yields dihydrosterculic acid (See Table 2.5). These reactions occur in a number of bacteria and in certain families of higher plants, Malvaceae and Sterculaceae.

When J.H. Law and his colleagues at Harvard University purified cyclopropane 'synthetase' from Clostridium butyricum, they found that the enzyme would catalyse the formation of cyclopropane fatty acids from ^{14}C-labelled methionine only if phospholipids were added in the form of micellar solutions (see p. 142 and chapter 6).

They discovered that the real acceptor for the methylene group was not the free monounsaturated fatty acid or its CoA or ACP thiol ester, but phosphatidyl ethanolamine − the major lipid of the organism. A pathway for cyclopropane fatty acid formation involving a phospho-

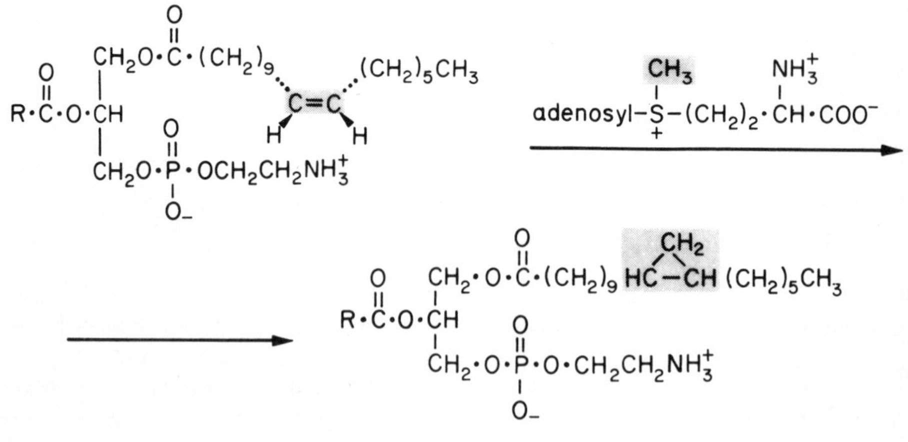

Fig. 2.14. Formation of cyclopropane fatty acid in Clostridium butyricum.

lipid has not been definitely proved in higher plants however.

The biosynthesis of cyclopropane and the related cyclopropene acids in higher plants has been studied by 'labelling experiments' with radioactive precursors. In this method, a supposed precursor for the compounds being studied is supplied to the plant and its uptake into various more complex molecules is measured at different time intervals. The sequence in which certain compounds are 'labelled' indicates the pathway by which those compounds are biosynthesized. Such experiments suggest that cyclopro*pene* acids are derived from cyclopro*pane* acids by desaturation. Suggestions that cyclopropenes arise by addition of a methylene group across an acetylenic bond have not been confirmed.

Pathways for the degradation of fatty acids

β–Oxidation

Long chain fatty acids, combined as triglycerides, provide the long term storage form of energy in the animal in the adipose tissue. The most important mechanism by which these storage fatty acids are degraded in a step-wise manner to yield energy which can be used in the form of 'energy-rich' molecules such as ATP, is known as β-oxidation — so called because the β or 3-carbon atom of the fatty acid chain is the site at which oxygen is introduced during the oxidation.

The mechanism was 'discovered' by chemical methods and intuition fifty years before the enzymology was understood.

The story of the discovery of β–oxidation is a classic case of the formulation of a correct hypothesis for a biological mechanism without any knowledge of the enzymic steps involved. In 1904, F. Knoop formulated a cyclic process for fatty acid degradation on the basis of feeding experiments with fatty acids which were 'labelled' not with ^{14}C as would be the case today, but with phenyl groups. The intermediate forms in Knoop's scheme were essentially the same as those identified 50 yr later by enzymological techniques.

There the matter rested until 1944 when interest in the enzymology of the process was stimulated by the experiments of L.F. Leloir and J.M. Munoz, which showed that fatty acid oxidation could take place in cell-free preparations of guinea pig liver. From there the reaction could be narrowed down to the *mitochondria* and from the cell particle, work moved over to the study of purified soluble enzyme preparations — a necessary step when so many of the intermediates are present in catalytic quantities only. From 1945, discoveries came thick and fast. The discovery of coenzyme A by F. Lipmann, the elucidation of its structure, the demonstration that it forms thiol esters with acetate and long chain fatty acids by Lynen and that these thiol esters were the 'active forms' of the respective acyl groups were important preliminaries in unravelling the β–oxidation pathway. No less important were the technical achievements such as large scale isolation of CoA by means of its cuprous salt and the enzymic synthesis of acyl-CoA esters by the team at the Enzyme

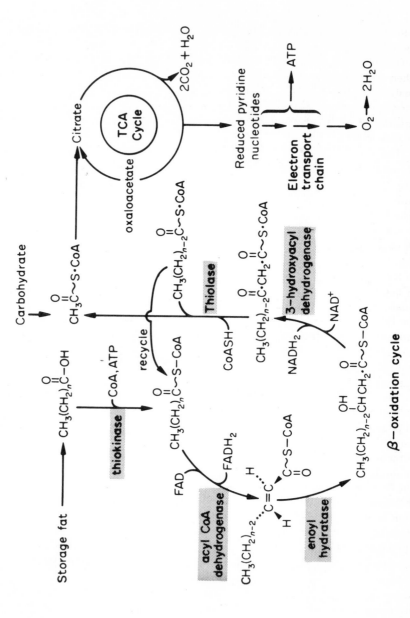

Fig. 2.15. β-Oxidation and energy metabolism.

Institute, Wisconsin. Wieland's method for the chemical synthesis of acyl-CoA derivatives was developed slightly later when most of the enzymes of the sequence had already been studied, although Lynen used chemically synthesized 'model compounds' to study three of the four enzymes of the cycle.

The sequence of reactions of the β–oxidation cycle is illustrated in Fig. 2.15 and a description of the individual steps is given in Table 2.13. The fatty acid must first be activated to its acyl-CoA thiol ester as we discussed on page 31. A point to remember here is that fatty acid activation is not restricted to mitochondria; indeed, activating enzymes of one sort or another have been found in most compartments of the cell. However, the four enzymes of the β–oxidation cycle proper are, in the animal kingdom at least, *strictly* mitochondrial. Although the vast majority of research into β–oxidation has been done with animal mitochondria, nevertheless the pathway is not unique to the animal kingdom but has also been found in peanut mitochondria and in tubercle bacilli. The only well documented non-mitochondrial system is that of the anaerobic micro-organism, *C. kluyveri*, but there the pathway is more geared for the reverse direction — synthesis, rather than breakdown. In higher plants, β–oxidation is closely geared to the glyoxylate cycle as a means by which the plant converts fats into sugar. Recent evidence suggests that the enzymes of these two important metabolic cycles are present in particles, quite distinct from mitochondria, which have been called variously, *glyoxysomes*, *peroxysomes* or *microbodies*.

Enzymes of β–oxidation

The first step in the β–oxidation cycle proper is the introduction of a *trans–αβ–* double bond into the hydrocarbon chain of the activated fatty acid, catalysed by the flavoprotein enzyme, *acyl–CoA dehydrogenase.*

The acyl-CoA dehydrogenases are unique among flavoproteins for a number of reasons. The binding of substrate and enzyme is so tight as to be virtually non-dissociable. The flavin moiety is very stable and not oxidized in the presence of molecular oxygen, ferricyanide or other electron acceptors. For this reason, the oxidation of the reduced form of the enzyme must be catalysed by a specific flavoprotein — the electron transport flavoprotein (ETF) discovered in 1956 by the American biochemists Crane and Beinert. This seems to be a unique case where an oxidation-reduction enzyme requires another specific protein for its reoxidation.

Enoyl–CoA hydratase

The hydratase catalyses the addition of the elements of water across the *trans* double bond of the unsaturated acyl CoA to form 3-hydroxyacyl–CoA (β-hydroxyacyl-CoA). It is fairly well established that a *trans* double bond in the unsaturated acyl–CoA is transformed into *L*(+)–3–hydroxyacyl–CoA (the hydroxy acid having the *absolute* configuration of the *L*-series and positive optical rotation). In the past, controversies about the absolute configuration of this intermediate arose because no less than 5 enzymic activities are possessed by the same highly purified crystalline enzyme. (See Table 2.13).

Table 2.13– Enzymes of the β–oxidation cycle (The reaction scheme is given in Fig. 2.14).

Enzyme		Description	Assay
Full name	Trivial name		
1 Acyl–CoA dehydrogenase (EC. 1·3·2·2)		Introduces a *trans*-2,3-double bond into the activated acyl chain. Chain-length specificities: 'green flavoprotein': $C_4 - C_8$ 'yellow flavoprotein' (medium chain) : $C_8 - C_{12}$ 'yellow flavoprotein' (long chain) : $C_8 - C_{18}$ All have 2 moles FAD/mole protein. The green F.P. has no copper; colour probably due to a complex of the prosthetic group with a chemical group of the protein.	Reversible reaction. (i) Couple reaction to reduction of electron acceptor in presence of E.T.F. (ii) Measure reduction of un-saturated acyl-CoA (reverse reaction) by certain dyes.
2 Enoyl–CoA Hydratase (EC 4.2.1.17)	Enoyl hydrase or Crotonase	Single protein active on $C_4 - C_{18}$. Has been purified to homogeneity and crystallized. Hydrates *trans*-2-enoyl-CoA to $L(+)$–3–hydroxyacyl–CoA but also has 5 other activities: (i) Hydratase: *cis*–2–enoyl–CoA $\rightleftharpoons$ $D(-)$–3–hydroxyacyl–CoA (ii) positional isomerase ($\Delta 2 \rightleftharpoons \Delta 3$) (iii) *cis-trans* isomerase (iv) transenoylase (enoyl-S-pantetheine + CoA $\rightleftharpoons$ enoyl-S-CoA + pantetheine) (v) Racemase $D(-)$–3–OH $\rightleftharpoons L(+)$–3–OH	Measure : (i) change in absorption of $\Delta 2$-unsaturated ($\alpha\beta$–unsaturated) bond at 232nm (ii) Couple the reaction with the next enzyme of the sequence (3–hydroxyacyl–CoA de-hydrogenase) and measure change in absorption of $NAD^+/NADH$ at 340nm

	Enzyme		Description	Assay
	Full name	Trivial name		
3	L–3–hydroxy–acyl– CoA dehydrogenase (EC 1.1.1.30)		Converts $L(+)$–3–hydroxyacyl–CoA into 3–Ketoacyl–CoA. Completely specific for L–configuration and NAD^+. Less specificity for chain length or CoA. An example of an enzyme which can be studied with Lynen model compounds. At pH7 equilibrium favours OH acid; at high pH equilibrium shifts in favour of Keto acid. 100% yield by trapping Keto acid with Mg^{++}.	Measure appearance or disappearance of NADH absorption at 340 nm.
4	Acetyl-acyl-CoA transacetylase E.C.2.3.1.9	Thiolase	Catalyses a thiolytic cleavage of the Keto acid in which –SH group of CoA displaces an acetyl–CoA moiety. One enzyme (purified and crystallized) for $C_4 \rightarrow C_{18}$. Purified protein catalyses 2 reactions: Thiolytic cleavage: $R \cdot CH_2 \cdot CO \cdot CH_2 S \cdot CoA + HS{-}E \rightleftharpoons$ $R \cdot CH_2 \cdot CO \cdot S{-}E + CH_3 COS \cdot CoA$ Acyl transfer: $R \cdot CH_2 \cdot CO \cdot S{-}E + CoASH \rightleftharpoons R \cdot CH_2 \cdot CO \cdot S \cdot CoA +$ $HS{-}E$ Overall thiolase:— $R \cdot CH_2 \cdot CO \cdot CH_2 COS \cdot CoA + COASH \rightleftharpoons$ $R \cdot CH_2 \cdot COS{-}CoA + CH_3 COS \cdot CoA$	(i) Measure appearance or disappearance of CoA groups by nitroprusside reaction. (ii) Measure appearance or disappearance of acetoacetyl–CoA band at 303 nm (iii) Link reaction to citrate synthesis in presence of acyl–CoA, malate dehydrogenase and condensing enzyme. Measure change in adsorption of NADH at 340 nm.

L—3—hydroxyacyl—CoA dehydrogenase

The next dehydrogenation step, which converts the L-hydroxy-compound into the corresponding keto-compound, is completely specific for the L-absolute configuration and for NAD^+ as cofactor, but most preparations also contain an activity specific for the D—configuration which tends to give the overall appearance of a *racemase*.

Acetyl-acyl-CoA transacetylase

The last enzyme of the sequence is generally known as *thiolase* and catalyses a *thiolytic cleavage* of the keto acid in which the —SH group of CoA displaces an acetyl—CoA moiety (see Fig. 2.15). The remaining fragment is thus an acyl—CoA two carbon atoms shorter than the molecule which began the cycle and the process can be repeated over and over again until the carbon chain has been completely oxidized to C_2 fragments. Lynen has recently (1968) obtained highly purified and crystalline thiolase and studied its properties. One enzyme appears to catalyse the cleavage of the whole range of keto acids from C_4 to C_{18}. The protein has a molecular weight of about 170,000 but in the presence of 5M urea dissociates reversibly into 4 sub-units of M.W. about 42,000. By incubating the enzyme with ^{14}C—acetyl—CoA, Lynen could demonstate that 3 molecules of acetate were bound per molecule of protein. The substance iodacetamide (which prevents binding of substrates to protein —SH groups) was able to completely inactivate substrate binding also when 3 moles were incorporated per mole of protein. Although this suggests the presence of 3 binding sites, the German group believe that a *completely* active enzyme would have four sites, in other words, one binding site per sub-unit. Of particular interest was the finding that the purified enzyme possesses two activities which could be demonstrated separately (see Table 2.13).

Other fatty acids containing branched chains, double bonds and an odd number of carbon atoms can also be oxidized.

So far, we have assumed that the fatty acid being oxidized is a straight chain, fully saturated compound. This is not necessarily the case and the ease with which other compounds are oxidized depends on the position along the chain of the 'extra group' or the capacity of the cell for dealing with the end products. From acids of odd chain length, one of the products is propionic acid and the ease with which the organism oxidizes the fatty acid is governed by its ability to oxidize propionate. Liver, for example is equipped to oxidize propionate and therefore deals with odd-chain acids quite easily; heart, on the other hand, cannot perform propionate oxidation and degradation of odd-chain acids grinds to a halt. The end product of propionate oxidation, succinyl-CoA, arises by a mechanism involving the B12 coenzyme : —

biotin
propionyl-CoA
carboxylase

$\boxed{B\,12}$

$$CH_3CH_2COSCoA + CO_2 + ATP \rightleftharpoons CH_3CH(COOH)COSCoA + ADP + P_i \rightleftharpoons COOHCH_2CH_2COSCoA$$

methylmalonyl-CoA · · · · · · · · · succinyl-CoA

Similarly, branched chain fatty acids with an even number of carbon atoms may eventually yield propionate, while the oxidation of the odd numbered acids proceeds by a different route involving 3-hydroxy-3-methylglutaryl-CoA (HMG—CoA):

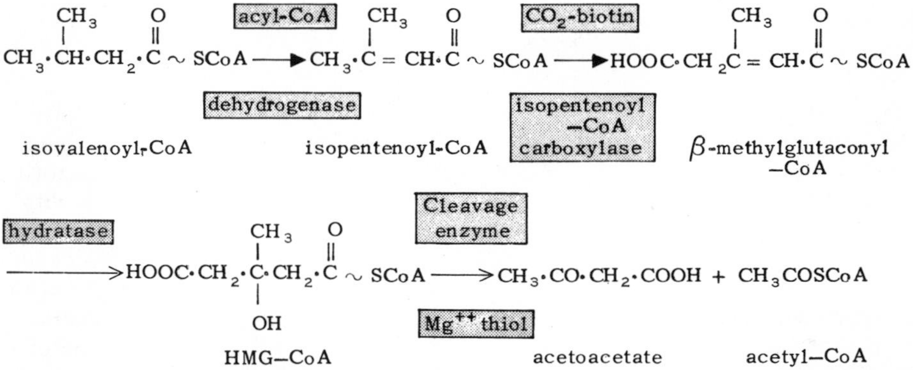

CoA. Once over this obstacle, β—oxidation can again continue to eliminate a further four carbon atoms. The $cis-\Delta^{12}$ double bond of the original unsaturated fatty acid now emerges as the $cis-\Delta^2$ bond of the resulting eight-carbon acid. In a series of papers pub-

$$\underset{\text{isovalenoyl-CoA}}{\overset{CH_3 \quad\quad O}{\underset{|}{CH_3 \cdot CH \cdot CH_2 \cdot C} \sim SCoA}} \xrightarrow[\text{dehydrogenase}]{\boxed{\text{acyl-CoA}}} \underset{\text{isopentenoyl-CoA}}{\overset{CH_3 \quad O}{CH_3 \cdot C = CH \cdot C \sim SCoA}} \xrightarrow[\substack{\text{isopentenoyl}\\ -CoA \\ \text{carboxylase}}]{\boxed{CO_2\text{-biotin}}} \underset{\substack{\beta\text{-methylglutaconyl}\\ -CoA}}{\overset{CH_3 \quad O}{HOOC \cdot CH_2C = CH \cdot C \sim SCoA}}$$

$$\boxed{\text{hydratase}} \xrightarrow{\quad} \underset{\text{HMG—CoA}}{\overset{CH_3 \quad\, O}{\underset{OH}{HOOC \cdot CH_2 \cdot C \cdot CH_2 \cdot C \sim SCoA}}} \xrightarrow[\boxed{Mg^{++}\text{thiol}}]{\boxed{\substack{\text{Cleavage}\\\text{enzyme}}}} \underset{\text{acetoacetate}}{CH_3 \cdot CO \cdot CH_2 \cdot COOH} + \underset{\text{acetyl—CoA}}{CH_3COSCoA}$$

Two problems arise during the oxidation of an unsaturated fatty acid, such as linoleic acid, ($\Delta^{9,12}C_{18:2}$). The first three cycles of β—oxidation can proceed normally enough to eliminate carbon atoms 1—6. At this stage we have a fatty acid with a cis-Δ^3 double bond, whereas the β—oxidation cycle requires the $trans$-Δ^2—isomer. The German biochemist W. Stoffel, has shown that an isomerase exists to convert the cis-Δ^3 compound into the necessary $trans$-Δ^2 fatty acyl-

lished in *Hoppe-Seyler's Zeitschrift* Stoffel described the isolation of a $cis-\Delta^2$ enoyl hydratase which catalyses the formation of the D(–)–3–hydroxyacyl—CoA. An *epimerase* (or *racemase*) then transforms this somewhat unusual isomer into the more normal L(+)–3–hydroxyacyl—CoA so that β—oxidation can continue unhindered. These modifications of the normal β—oxidation sequence to accommodate cis-unsaturated fatty acids are illustrated in Fig. 2.16.

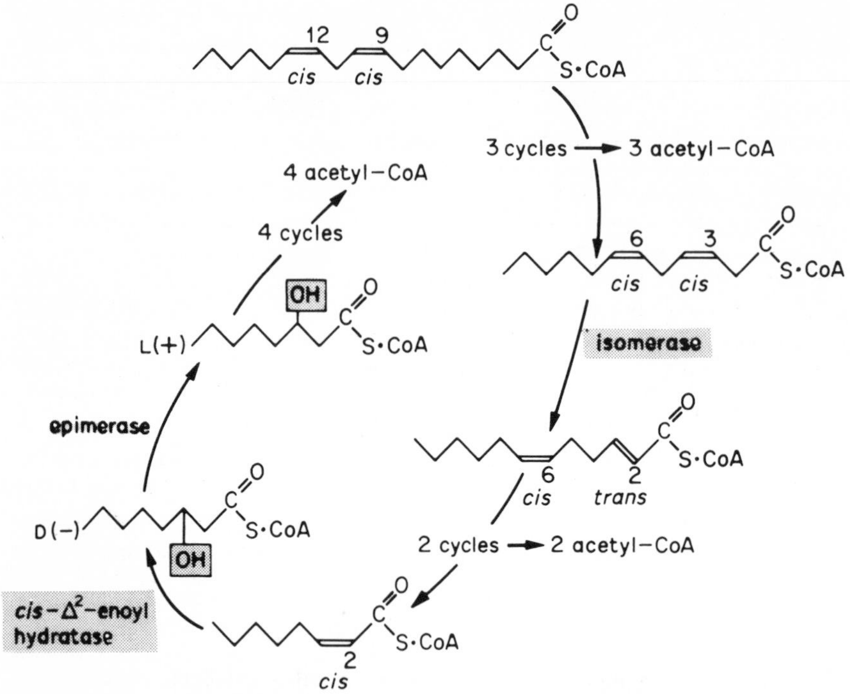

Fig. 2.16. β—Oxidation of linoleic acid.

*Ketone Bodies' are a feature
of liver fatty acid oxidation.*

Our recent discussion of the β—oxidation cycle has implied that the last keto acid of the sequence, namely acetoacetate, is cleaved in the normal way by the thiolase enzyme to yield finally two molecules of

acetoacetic acid has been formed, liver lacks the thiokinase to reconvert it into the CoA ester. The formation of the free acid may be due to the presence of a deacylase or by a quite different pathway involving the formation and finally the cleavage of HMG—CoA:

$$\text{CH}_3\cdot\text{CO}\cdot\text{CH}_2\text{COSCoA} + \text{CH}_3\text{COSCoA} \longrightarrow \begin{matrix} \text{CH}_2\text{COSCoA} \\ | \\ \text{CH}_3\text{C}\cdot\text{OH}\cdot\text{CH}_2\text{COOH} + \text{CoA} \end{matrix}$$

$$\longrightarrow \text{CH}_3\text{CO}\cdot\text{CH}_2\text{COOH} + \text{CH}_3\text{COSCoA}$$

acetyl—CoA. This is the case in most tissues, but the liver differs in that it accumulates acetoacetate and various other ketones. It appears that once

HMG is an important intermediate in cholesterol biosynthesis and hence the presence of this pathway should not be regarded as one which 'wastes' acetyl—CoA but viewed as a link between the pathways of fatty acid metabolism and cholesterol biosynthesis.

Ketone bodies also accumulate in body fluids during certain physiological and pathological states when the rate of carbohydrate metabolism is low. Thus, when the rate of production of acetyl–CoA by β–oxidation exceeds the rate at which it can condense with oxaloacetate or be otherwise utilized, then an excessive accumulation (*ketosis*) of ketone bodies will occur.

How β–oxidation is linked with other mitochondrial processes for the production of useful energy.

Until they are ready to be consumed by β–oxidation, fatty acids are stored in the triglycerides of the adipose tissue and 'mobilized' when required by mechanisms which will be discussed in chapter 3. We have yet to see how they arrive at the oxidation site and how the final products are integrated into the overall processes of mitochondrial metabolism to provide useful energy.

How they arrive is an important point because it turns out that acyl–CoA derivatives are not able to penetrate the mitochondrial membrane. Recently, it has been possible, by a variety of techniques to separate the inner and outer membranes of the mitochondrion. It seems likely that the enzyme systems concerned with the TCA cycle, the electron transport chain and β–oxidation are bound to the inner membrane. However, owing to the different techniques used by different research teams, there is considerable controversy about this at the present

time and we must wait for methods to be standardized and stand the test of time before making a categoric statement. Nevertheless, some kind of transport through a barrier is undoubtedly necessary.

Role of carnitine

During the 1960's, the American, Fritz, and the Norwegian, Bremer, and others have elaborated a theory in which it is postulated that the membrane is permeable to long chain fatty esters of the base (–)carnitine and that it is in

$$(CH_3)_3 \overset{+}{N} \cdot CH_2 \cdot \underset{\underset{OH}{|}}{CH} \cdot CH_2 \cdot COO^-$$

carnitine

this form that long chain fatty acids are transported into the mitochondria to the site of oxidation. The ideas began with the finding that addition of carnitine stimulated the oxidation of fatty acids *in vitro*; acyl-carnitines were oxidized even more rapidly. The theory really began to take shape with the discovery of the enzyme *carnitine-acyl transferase* which transfers the long chain acyl group from CoA to carnitine. Once inside the membrane, the reverse reaction can take place and the CoA ester is delivered to the site of β–oxidation. In a truly 'soluble' system, the stimulatory effect of carnitine is abolished because no impermeable barrier exists. These ideas are illustrated in Fig. 2.4. It is only fair to say that at the time of writing this book, several scientists have reported that some mitochondria can oxidize fatty acids without the need for carnitine;

indeed in the flight muscle mitochondria of some moths, which oxidize palmitate at 20 times the rate of mammalian mitochondria, *carnitine-palmitoyl transferase* could not be detected. In this case the fatty acid may be transported as free fatty acid. The overall picture is therefore extremely confused and it is likely to be several years before the role of carnitine-dependent and carnitine-independent systems can be assessed. A separate enzyme, *carnitine-acetyl transferase*, is quite widespread and, as we have seen, may function to transport acetyl—CoA from the mitochondria to extramitochondrial sites of fatty acid synthesis. The role of carnitine may therefore not be restricted to the area of β—oxidation.

The production of acetyl—CoA as the end product of β—oxidation does not of itself furnish the cell with useful energy — it merely provides it with an important metabolic intermediate. The importance of β—oxidation as an energy source lies in the way in which acetyl—CoA is further metabolized by mitochondrial reactions. Acetyl—CoA is also an end product, via glycolysis, of carbohydrate breakdown and thereby provides a link between carbohydrate and lipid catabolism and an entry into what Mahler has called 'the hub of the metabolism of almost all cells' — namely the tricarboxylic acid cycle (TCA cycle). The cycle is initiated by the condensation of acetyl—CoA with oxaloacetate (Fig. 2.15) a catalytic amount of which is needed to prime or 'spark' the oxidation of a relatively large amount of acetyl—CoA to CO_2 and water. Neither carbohydrate oxidation nor β—oxidation can function without this sparking phenomenon, but whereas oxaloacetate can be formed directly from carbohydrates by carboxylation of pyruvate, no such possibility arises from fatty acids and hence the operation of the β—oxidation cycle is entirely dependent on the 'sparker'. The energy picture is completed by the oxidation of the reduced coenzymes, generated by the TCA cycle, in the electron transport chain and the generation of ATP by the coupling of electron transport with oxidative phosphorylation.

α—Oxidation

β—oxidation is not the only mechanism for fatty acid degradation.

While it is generally true that β—oxidation is quantitatively the most important method by which fatty acids are biochemically degraded, there are nevertheless several other oxidation pathways. One of the alternatives is α-oxidation, so called because only *one* carbon atom, the carboxyl-carbon is lost at each step and carbon atom two (the α-carbon) becomes oxidized to the new carboxyl group. Although the pathway was first observed in plants and subsequently plant material has most often been used in studying the mechanism of α-oxidation, it also occurs in mammalian tissues, mainly in brain.

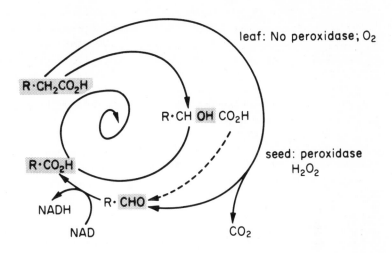

leaf: No peroxidase; O_2

$R \cdot CH_2CO_2H$

$R \cdot CH$ OH CO_2H

seed: peroxidase
H_2O_2

$R \cdot CO_2H$

$R \cdot CHO$

NADH

NAD

CO_2

Fig. 2.17. Pathways for α—oxidation in leaf and seed.

More than one α-oxidation mechanism may exist.

α—Oxidation has received nothing like the detailed attention which has been given to β—oxidation and it is difficult to give a simple description of the pathway because the available data are conflicting and point to the possibility of more than one mechanism. The pathways worked out by the Unilever Research team in England for the oxidation of fatty acids in *leaf tissue* are shown in Fig. 2.17. By using as substrates, fatty acids which were labelled in all carbon atoms ('uniformly labelled'), each intermediate in the pathway became labelled and could be studied when the products were separated by gas—liquid radio-chromatography. Alternatively, the overall oxidation could be assayed by measuring the loss of ^{14}C from carboxyl-labelled substrates. Oxygen and NAD were the only cofactors required to produce the next lower acid in the homologous series, with α—hydroxy acids as intermediates in their formation. If NAD were omitted, the next lower aldehyde accumulated. The original work defining the pathway of α—oxidation in plants was done by P.K. Stumpf and his team in California. They studied the pathway in the cotyledons of germinating peanuts. Again NAD was required to oxidize an aldehyde intermediate but the aldehyde could only be formed — in contrast with the leaf system — if a source of hydrogen peroxide and a peroxidase enzyme were

present. The pathways will have to be
'dissected' more carefully before it can
be categorically stated that α-oxidation
in leaves and seeds are two different
processes.

In brain, α-hydroxy acids occur as
important components of the *cerebrosides* –
glycolipids which accumulate gradually
as the brain ages (chapter 5). J.F. Mead's
research team at the University of
California have isolated brain subcellular
particles which catalyse the decarboxylation
of α-hydroxy acids to yield the next
lower 'normal' acid of the homologous
series, probably by way of an enzyme-
bound keto-acid intermediate.

The formation of hydroxy acids
is stereospecific. The substrates
do not require 'activation'.

Like most enzymic reactions, the formation
of the hydroxy acid in α-oxidation is
stereospecific in that only the L-hydroxy
acid is an *intermediate* in the oxidation
pathway. α-Oxidation contrasts with
β-oxidation however, in two important
respects. First, the fatty acid does
not require to be 'activated' – the non-
esterified fatty acid is a good substrate.
Second, the oxidation is not linked to
energy production as far as we know.
This raises the question of its function,
about which little is known. It may be
useful as a disposal mechanism especially
for degrading the very long chain fatty
acids which are found in brain and which
are not very readily transported or easily
handled by the more usual β-oxidation
system. In addition, α-oxidation has been
shown to occur in animal tissues

in conjunction with β-oxidation in for
example the degradation of a methyl branch
chain fatty acid in which the methyl group
is β to the carboxyl group (i.e. in position
3). In this case, β-oxidation is blocked
but can proceed normally if the obstacle
is removed by an initial α-oxidation.

Refsum's disease

This mechanism has significance in
connection with *Refsum's disease*, an
hereditary disease characterized by an
accumulation of 3,7,11,15-tetramethyl
hexadecanoic acid (phytanic acid) which
arises from the phytol present in dietary
vegetables. Phytanic acid cannot be
degraded by β-oxidation because of the
3-methyl group. Therefore an initial
α-oxidation is necessary, followed by
β-oxidation. Healthy people can meta-
bolize phytanic acid, but patients with
the disease lack the enzymes of
α-oxidation and so accumulate the acid
in their tissues. The disease is
characterized by chronic polyneuropathy,
night blindness, narrowing of the visual
field, skeletal malformation and cardiac
involvement and is normally fatal.

ω-oxidation

ω-oxidation is catalysed by
microsomal enzymes of the
'oxygenase' type. Its function
is uncertain.

ω-oxidation contrasts with α- and
β-oxidation in that the oxidative attack
is remote from the carboxyl group and
results in an ω or ω-1 hydroxy acid and

ultimately in a dicarboxylic acid. There is also a marked difference in subcellular location. Whereas β—oxidation is an exclusively mitochondrial, energy producing process, which is capable of complete degradation of a fatty acid, ω—oxidation enzymes are located in the endoplasmic reticulum and the process inevitably stops at the formation of the dicarboxylic acid. The location of α-oxidation is less clearcut; there is evidence for components of α-oxidation in mitochondria, endoplasmic reticulum, and in the cell sap.

The properties of the enzymes involved in ω—oxidation have been studied by a Japanese research team under M. Kusunose and by M.J. Coon in America. The cofactors needed, namely NADH, NADPH, molecular oxygen, and ferrous ions, and the fact that hydroxy compounds are formed in the reaction suggested very strongly that one of the enzymes was of a type known as a *mixed function oxygenase*. In oxidations of this type, one atom of the oxygen molecule is transferred to the substrate while the other is reduced by the pyridine nucleotide coenzyme — hence the name *mixed function*. Such enzymes are characteristic of the endoplasmic reticulum where they not only catalyse the oxidation of fatty acids but of steroids, alkanes or aromatic compounds — an important means of destroying toxic compounds and drugs. The complete oxidation requires several protein components which differ from one tissue or organism to another. Apart from the *hydroxylase* itself, the pathway involves a number of electron carriers — the cytochromes — by means of which electrons are transferred from the reductant, NADPH, to oxygen to form water. A simplified scheme is illustrated in Fig. 2.18.

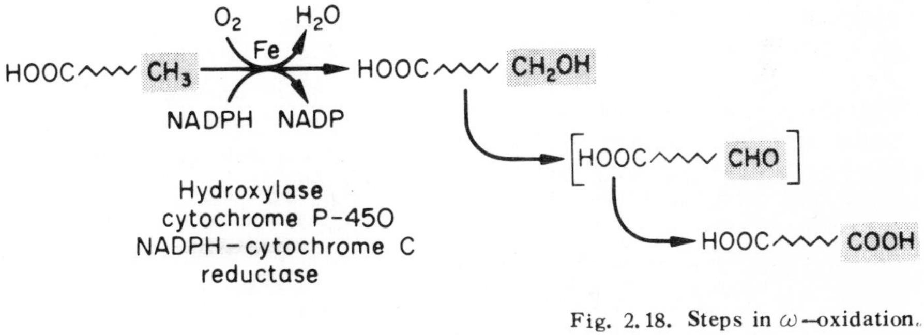

Fig. 2.18. Steps in ω—oxidation.

Peroxidation

Lipid peroxides are formed by several mechanisms: by autoxidation, by chemical catalysis with haematin or by enzymic catalysis with lipoxidase.

One of the characteristic reactions of lipids which are exposed to oxygen is the formation of peroxides. This phenomenon is of immense practical importance for it leads both to the deterioration of tissues *in vivo* and is also responsible for the spoilage of foods. One of the limiting factors in the storage of even freeze-dried foods is damage caused by lipid peroxides. This is due to the free radicals produced during peroxide formation, such as ROO˙, RO˙, OH˙; these react at random by hydrogen abstraction and a variety of addition reactions to damage enzymes, proteins, other lipids and vitamins. Vitamin A is particularly susceptible to peroxidation damage. Compounds which react quickly with free radicals — *antioxidants* such as the tocopherols — are useful in retarding peroxidation damage. An example of a built-in biological antioxidant is vitamin E; animals which are deficient in vitamin E are particularly susceptible to tissue damage by peroxidation.

Peroxidation may occur by *autoxidation*; in other words, the lipid catalyses its own oxidation. It is the unsaturated acids which are susceptible to oxidation of this type; the more highly unsaturated the acid, the greater is its susceptibility to chemical oxidation. It is probable, however, that the importance of autoxidation has been overemphasized and that in many cases the reaction is catalysed by haematin compounds, haemoglobin, myoglobin or the cytochromes. Although the products are similar in each case, the two types of oxidation can be distinguished by the fact that autoxidation is much slower, has a very much greater activation energy and is not inhibited by cyanide. We should emphasize that both these types of oxidation are chemical and not enzymic.

Peroxidation catalysed by lipoxidase is a much more specific process.

The third kind of peroxidation which we will describe is catalysed by the enzyme *lipoxidase* (or *lipoxygenase*). So far the enzyme has only unequivocally been found in plants. Claims for its presence in animal tissues are confused by the widespread occurrence of haematin catalysis. The chief sources of the enzyme are in peas and beans (especially soy bean), cereal grains and oil seeds. It was originally detected by its oxidation of carotene and has been used extensively in the baking industry for bleaching carotenoids in dough.

Lipoxidase catalyses the reaction:

$$R \cdot CH = CH - CH_2 - CH = CH \cdot R_1 + O_2 \longrightarrow R \cdot CH = CH - CH = CH - \underset{\underset{OH}{\overset{\mid}{\underset{\mid}{O}}}}{CH} - R_1$$

cis   *cis*   *cis*   *trans*

When H. Theorell and his colleagues in Sweden purified and crystallized soy bean lipoxidase in 1947, they revealed a fact which makes this enzyme unique among oxidation enzymes — it has no prosthetic group or heavy metal associated with it; nor is there an 'active —SH group' for the enzyme is not inhibited by any of the classical —SH poisons such as p-chloromercuribenzoate or iodoacetate. The product of the enzymic reaction — a hydroperoxide — is similar to the products of purely chemical catalysis but the lipoxidase reaction has a number of distinguishing features. The activation

energy is smaller than that for chemical reactions, and the *enzyme has very specific substrate requirements.* In order to be a substrate, the fatty acid must contain at least two *cis* double bonds interrupted by a methylene group. Linoleic acid is the best known substrate. Most of the acids which have been tested and which show high reaction rates, have a double bond six carbon atoms from the methyl end of the chain. The distance of the carboxyl group from the diene system is less critical but it must not be sterically hindered, indicating that attachment to the enzyme surface may occur at the

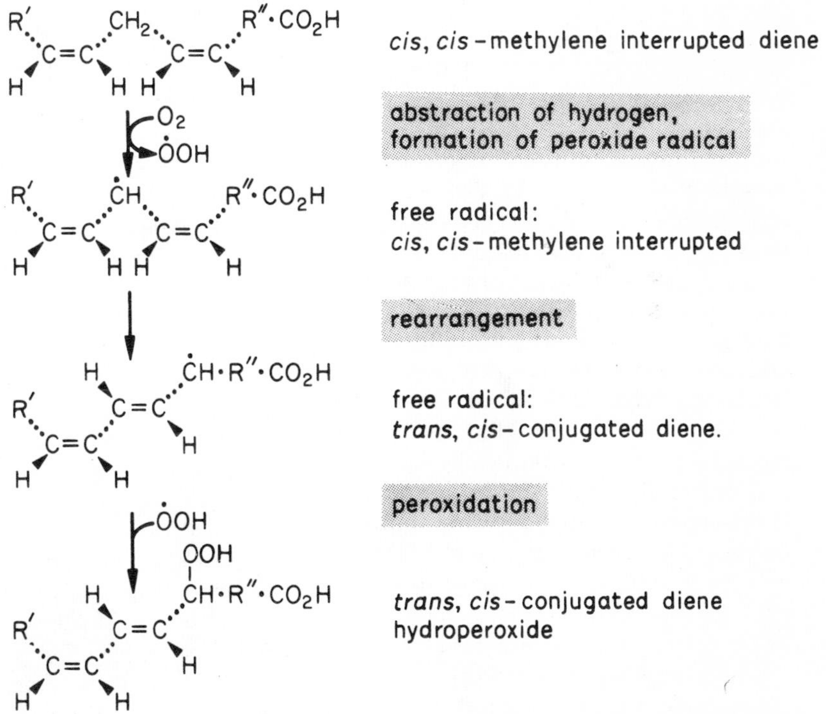

Fig. 2.19. Possible mechanism of lipoxidase.

carboxyl end of the chain. Like chemically catalysed peroxidation, the lipoxidase reaction involves free radicals and can be inhibited by radical trapping reagents such as the tocopherols. The reaction sequence shown in Fig. 2.19 has not been proved, but represents the most likely pathway.

The insatiable urge of biochemists to assign a function to every molecule in the living cell cannot yet be satisfied in the case of lipoxidase!

Quantitative and qualitative fatty acid analysis

As already pointed out chromatographic methods can not only separate closely related compounds but also help define their structure and in combination with suitable detection techniques give quantitative data as well. So far as fatty acids are concerned, it is only infrequently necessary to analyse mixtures containing all the fatty acids from C_1 to C_{20}; more usually one has the range C_2 to C_{10} and C_{10} to C_{20}. The two ranges are best analysed separately since the shorter chain acids are preferably handled and analysed as the free acids whereas the larger chain acids are best converted into their methyl esters. Esterification can be done either, (1) directly from the isolated lipid by refluxing with methanol-HC1 or methanol-sulphuric acid followed by extraction of the esters with light petroleum or (2) after saponification and extraction of the free acids with ether, by treatment with diazomethane in ether.

The short chain free acids (C_1 to C_8) can be separated on the gas chromatogram by 4 ft columns of dioctyl sebacate containing 15% sebacic acid at 150 °C or poly-diethyleneglycol-adipate at 125°C.

Mixtures of long chain fatty acids are more complex because of the occurrence of mono, di, tri and tetraenoic acids, particularly of C_{16}, C_{18} and C_{20} chain lengths. No single stationary phase is capable of separating every acid. Two different types of column are necessary to define the complexity of the mixture and a decision can then be made as to which column gives the necessary resolution.

Columns having saturated paraffin hydrocarbons (Apiezon L grease) or silicone greases separate largely on the basis of molecular weight; unsaturated acids emerge from the column earlier than the corresponding saturated acids, i.e. 18:4 before 18:3 + 18:2 (overlap), then 18:1, then 18:0 (See Fig. 2.20). There is some separation of isomeric unsaturated acids, and by the use of long capillary columns 11-octadecenoic acid can be distinguished from 9-octadecenoic acid, but better separations can usually be achieved by TLC on silica impregnated with silver nitrate (see page 86 and Fig. 2.24). Branched chain acids emerge before the saturated straight chain acid of the same carbon number, the higher the degree of branching the greater the separation. Thus these acids will overlap with the unsaturated acids.

A similar length column having polyethylene glycol adipate or other heatstable polyester gives separations based on number of double bonds as well as on molecular weight (i.e. chain length). The order of emergence from this column will therefore be 16:0, 16:1, 16:2, 16:3, 16:4, 18:0, 18:1, 18:2, 18:3, 18:4, etc. (see Fig. 2.21). Branched chain acids may

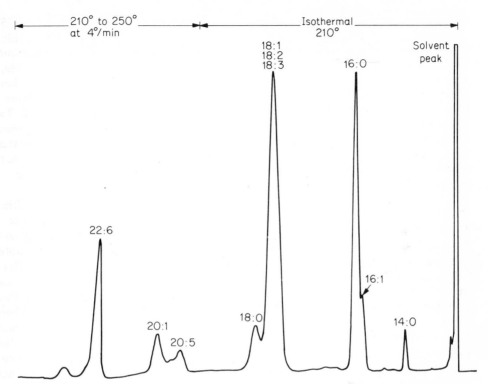

Fig. 2.20. Separation of fatty acid methyl esters by gas-liquid chromatography on a non-polar stationary phase. The fatty acid methyl esters were derived from trout flesh lipids. Stationary phase: SE–30; temperature: isothermal at 210° for 16 mins, then temperature programmed at 4° per min to a maximum of 250°.

therefore overlap highly unsaturated acids of shorter chain length.

Determination of the structure of an unknown fatty acid.

Even small amounts of a component can be isolated from the gaseous effluent from the gas chromatogram (provided the detector is in parallel with the outlet) by allowing the gas stream passage through a wide glass tube loosely packed with defatted cotton wool wetted with a solvent such as ether or methanol. Too

rapid cooling of the stream produces a very stable fog that cannot easily be condensed. Such an isolated component can be run on another type of column together with a reference compound to give a relative retention volume. Use of a graph of data derived from known fatty acids on two types of column as in Fig. 2.22 allows an unknown acid to be defined as straight or branched chain, saturated or unsaturated and also defines the degree of unsaturation. After isolation of an unsaturated acid it can be chemically

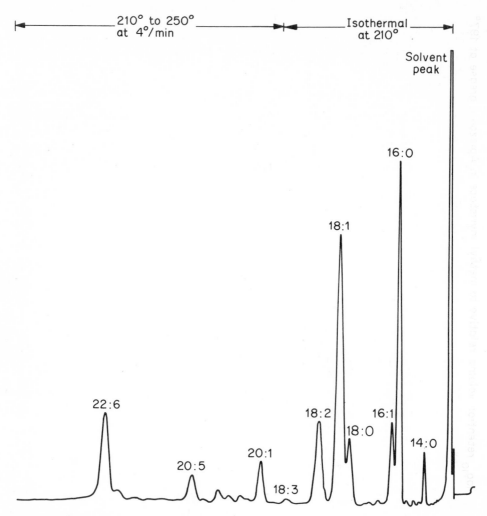

Fig. 2.21. Separation of fatty acid methyl esters by gas-liquid chromatography on a polar
stationary phase. The fatty acid methyl esters were derived from trout flesh lipids
Stationary phase: FFAP; temperature: isothermal at 210° for 13 mins, then temperature
programmed at 4° per min to a maximum of 250°.

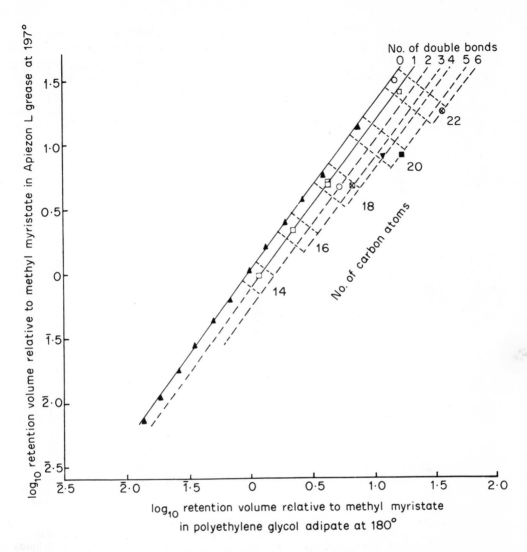

Fig. 2.22. Elucidation of the structure of a fatty acid from GLC data obtained from two different types of columns. Saturated long chain fatty acids, mono-, di- tri- tetra-, penta- and hexaunsaturated acids were chromatographed as methyl esters on two different columns. One has a polar stationary phase (polyethylene glycol adipate at 180°), the other a non-polar stationary phase (Apiezon L grease, 197°).

split at double bonds by permanganate or by ozonolysis to give mono and dicarboxylic acids that can be identified as their methyl esters or as half aldehydes by GLC. Final conclusions as to structure are however best made by use of the mass spectrometer, I.R. spectroscopy (for *trans* unsaturated acids) or N.M.R.

The area under the peak drawn by the potentiometric recorder is proportional to the mass of substance detected (if the detector has a linear response over the range required), though ideally each fatty acid needs to be run at a series of

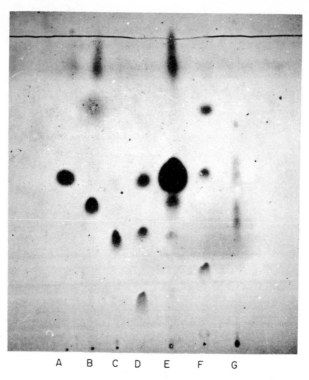

A   B   C   D   E   F   G

Fig. 2.23. Thin layer chromatography of fatty acid methyl esters and simple lipids on silica gel G impregnated with silver nitrate (5% w/w). Developing solvent: diethyl ether-petrol, 5:95 (v/v) and spots were located by spraying with 50% sulphuric acid and charring.

A. Methyl stearate.

B. Methyl elaidate.

C. Methyl oleate.

D. Methyl stearate, oleate and linoleate mixture.

E. Methyl esters from fecalith lipids.

F. Mixture of cholesterol stearate, cholesterol oleate and cholesterol linoleate.

G. Sperm oil (wax ester mixture).

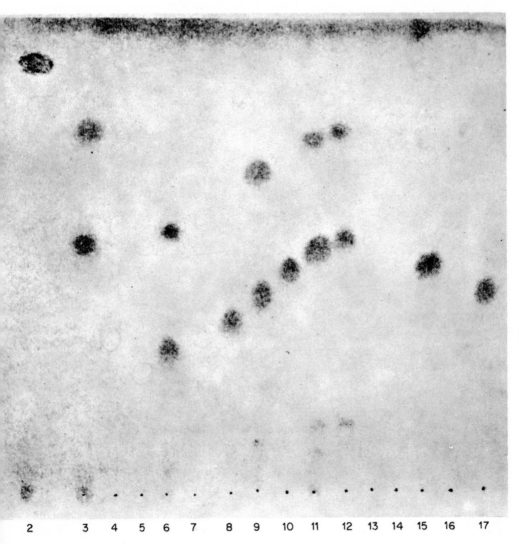

Fig. 2.24. Thin layer chromatography of isomeric octadecenoates on silver nitrate-silica gel G (30:70). The position of the double bond is indicated by the sample number, samples being 2,3,6,9,11 and 12 *trans*-octadecenoates, 3,6,8,9,10,11,12 and 15 *cis*-octadecenoates and the vinyl compound 17-octadecenoate. The plate was developed at −25°, three times with toluene. The spots were located with chlorosulphonic acid–acetic acid (1:2) and charring.

levels to determine the proportionality constant. The percentage of any component is then given by

$$\frac{\text{area of peak} \times 100}{\text{total area of all peaks}}$$

Thin layer chromatography

This is a very useful ancillary technique since inclusion of silver nitrate in the adsorbent allows selective retardation of unsaturated acids by reversible complexing with the Ag^+ ion. The greater the number of double bonds the greater the retardation (Fig. 2.23); *trans* acids are retarded less than the corresponding

cis acids, and positional isomers can also be separated in some cases (Fig. 2.24 Preparative TLC can be used to isolate groups of acids defined by the number of double bonds. Study of such a group by GLC and (after single component isolation), oxidative degradation followed by identification of the fragments by GLC, often allows the complete structure of a fatty acid to be defined.

Silver nitrate TLC is also useful for separation of mono-hydroxy acids. Dihydroxy or tetrahydroxy acids are best dealt with by TLC using borate or arsenate-impregnated silica gel.

SUMMARY

The fatty acids are long chain carboxylic acids, sparingly soluble in water. They form the common feature of *complex lipids* as defined in this book and are responsible for the hydrophobic properties of those compounds. The parent molecules are the long straight chain saturated acids but there may be many modifications or substitutions in the chain to produce *branched, unsaturated, hydroxy, keto, epoxy* or *cyclic* acids. The most abundant natural fatty acids are the *cis-mono-* or *poly-*unsaturated derivatives.

Polyunsaturated acids normally (but not always) have a methylene interrupted system of double bonds. Certain of these are essential for the good health of animals but cannot be made by them and so have to be supplied in the diet from plant sources. These *essential fatty acids* are converted by the animal into physiologically active cyclic, oxygenated, unsaturated fatty acids called *prostaglandins*.

Saturated fatty acids are formed by condensation of acetyl–CoA and malonyl–CoA with simultaneous elimination of CO_2 – essentially a stepwise condensation of 2-carbon fragments. Unsaturated acids are formed by the direct introduction of double bonds into the chain by an unknown mechanism involving *molecular oxygen* and *NADH* or *NADPH*. In anaerobic bacteria the *cis* double bond is introduced by isomerization of a *trans* bond during the early stages of the normal pathway of fatty acid biosynthesis. The organism's long chain unsaturated acids are then formed by elongation of the *cis*-unsaturated intermediate.

One of the chief *functions* of fatty acids is in the long term storage of metabolic energy in the form of triglycerides in adipose tissue. When required, the fatty acids are *mobilized* and oxidized in mitochondria by a stepwise cleavage of

2-carbon fragments (β–oxidation). The end product, acetyl–CoA, is further metabolized to yield ATP. Other oxidation mechanisms exist. These do not completely break down the fatty acid molecule but result in the formation of oxygenated or dicarboxylic acids.

Fatty acid biochemistry has been revolutionized mainly by two techniques — GLC and 'argentation' TLC which in combination allow almost complete resolution of fatty acid types with similar physical and chemical properties.

BIBLIOGRAPHY

Structure, chemical and physical properties, and distribution of fatty acids

1. GUNSTONE F.D. (1967). *An Introduction to the Chemistry and Biochemistry of fatty Acids and their Glycerides*, Chapman & Hall, London.

2. HILDITCH T.P. and WILLIAMS P.N. (1964). *The Chemical Constitution of Natural Fats*, Chapman & Hall, London.

3. MARKLEY K.S., (Ed.) (1967). *Fatty Acids, their Chemistry, Properties, Production and Uses*, 2nd ed., Interscience, New York.

4. CHAPMAN D. (1965). *The Structure of Lipids by Spectroscopic and X-ray Techniques*, Methuen, London.

Fatty Acid Biosynthesis

5. MAJERUS P.W. and VAGELOS P.R. (1967). *Fatty Acid Biosynthesis and the Role of the Acyl Carrier Protein. Advances in Lipid Research* 5, 1.

6. LYNEN F. (1967). The role of biotin — dependent carboxylations in biosynthetic reactions. *Biochem. J.* 102, 381.

7. STUMPF P.K. (1969). Fatty acid metabolism. *Ann. Rev. Biochem.* 38, 159. A good review of the most recent developments.

8. JAMES A.T. (1968). Biosynthesis of unsaturated acids by plants. *Chemistry in Britain* 4, 484.

9. BLOCH K. (1969). Enzymic synthesis of monounsaturated fatty acids. *Accounts of Chemical Research* 2, 193.

Essential Fatty Acids and Prostaglandins

10. ALFIN-SLATER R.B. and AFTERGOOD L. (1969). Essential fatty acids reinvestigated. *Physiol. Rev.* 48, 758.

11. BERGSTRÖM S. and SAMUELSSON B. (1968). The prostaglandins. *Endeavour* 27, 109.

Oxidation of fatty acids

12. MAHLER H.R. Biological oxidation of fatty acids. A chapter in reference 3, page 1487.

3 Neutral lipids: glycerides, sterol esters, vitamin A esters, waxes

Having described in detail the individual fatty acids which are found in nature esterified in complex lipids, we can now go on to discuss the different types of acyl lipids themselves.

Biochemists find it convenient to divide acyl lipids into two broad categories, *neutral* and *polar* lipids. Polar lipids are those which contain a polar group such as the phosphate group and 'base' of the phospholipids, the sulphate group of the sulpholipids or the sugar moiety of the glycolipids. These will be described in chapters 4 and 5. Neutral lipids contain no such areas of polarity. The most readily observable differences between the two classes in practical terms is in their physical properties and therefore in their solubility and chromatographic properties. Neutral lipids are much more readily soluble in completely apolar solvents such as hydrocarbons and during chromatography are much more easily eluted by these solvents than are polar lipids.

A definition based on physical properties and solubility, however, is necessarily very broad and inevitably this chapter will include a discussion of a range of compounds differing widely in chemical structure.

GLYCERIDES

Glycerides, or fatty acid esters of glycerol, are the major components of natural fats and oils.

Glycerides are esters of the trihydric alcohol, glycerol, and fatty acids. In triglycerides (triacylglycerols), all three glycerol hydroxy groups are esterified. Partial glycerides have only one or two positions esterified and are called mono- and diglycerides respectively (monoacyl- and diacylglycerols), (Fig. 3.1). Glycerides are the chief constituents of

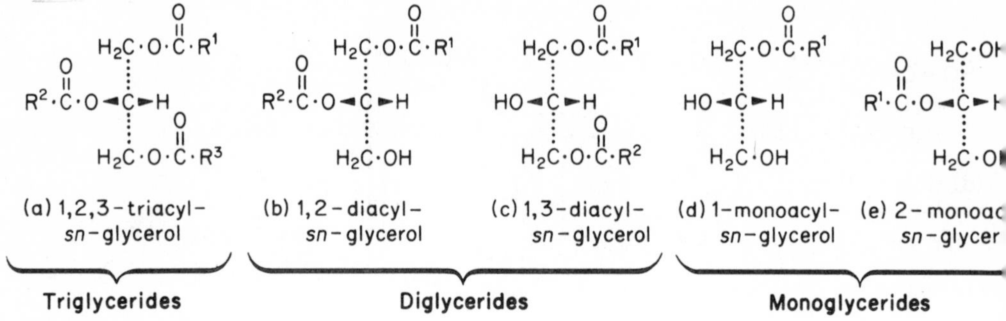

Fig. 3.1. Some common glyceride structures.

natural fats (solids) and oils (liquids), names which are often used synonymously for them, although it is important to remember that natural fats and oils also contain minor proportions of other lipids. The most abundant fatty acids in natural glycerides are palmitic, stearic, oleic and linoleic; plant glycerides have a relatively higher proportion of the more unsaturated acids. Milk fats have a much higher proportion of short chain fatty acids ($C_4 - C_{10}$). Odd chain or branched chain fatty acids are only minor constituents of glycerides; seed oils, especially, contain a variety of unusual fatty acids with oxygen-containing groups and ring systems, (see chapter 2). Some glycerides which contain ricinoleic acid (see chapter 2) may have further fatty acids esterified with the hydroxy group of the ricinoleic acid. Thus, tetra-, penta- and hexa-acid glycerides occur in some plant oils.

In recent years, interest has centered on glycerides containing acyl chains linked to glycerol by an ether rather than an ester linkage. Chimyl, batyl and selachyl alcohols are glyceryl monoethers derived from long chain alcohols corresponding to $C_{16:0}$, $C_{18:0}$, and $C_{18:1}$ respectively (Fig. 3.2a). These are the major naturally occurring glyceryl ethers and may be esterified in addition with one or two fatty acids (Fig. 3.2b,c). Usually both *alkyl* ethers (Fig. 3.2b) and *alkenyl* ethers (Fig. 3.2c) occur together in varying proportions, mainly in the liver oils of fish such as sharks. Only very small quantities of ether glycerides are found in plants.

The confirmation of the presence of diol lipids and the elucidation of their structure is also a fairly recent development: these compounds are found in small quantities in such widely differing tissues as mutton fat, fish liver, egg yolks, corn seeds, yeast and rat liver (Fig. 3.3).

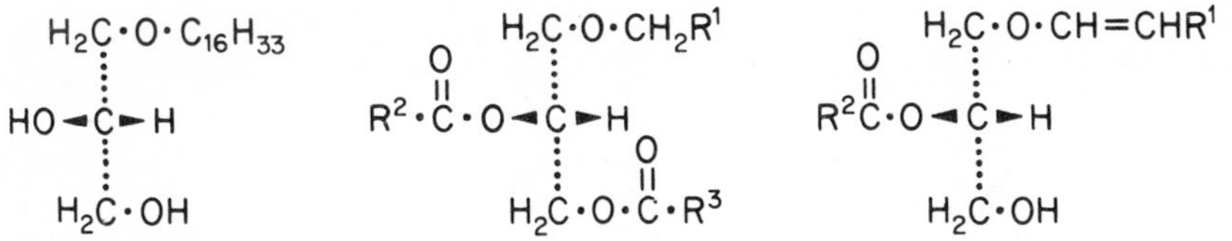

$$H_2C \cdot O \cdot C_{16}H_{33}$$
$$HO \blacktriangleleft C \blacktriangleright H$$
$$H_2C \cdot OH$$

(a) Chimyl alcohol
1-hexadecyl
– sn –glycerol

$$H_2C \cdot O \cdot CH_2R^1$$
$$R^2 \cdot \overset{O}{\underset{\|}{C}} \cdot O \blacktriangleleft C \blacktriangleright H$$
$$H_2C \cdot O \cdot \overset{O}{\underset{\|}{C}} \cdot R^3$$

(b) 1-alkyl-2,3-diacyl
– sn –glycerol

$$H_2C \cdot O \cdot CH = CHR^1$$
$$R^2\overset{O}{\underset{\|}{C}} \cdot O \blacktriangleleft C \blacktriangleright H$$
$$H_2C \cdot OH$$

(c) 1-alkenyl-2-acyl
– sn –glycerol

Fig. 3.2. Fatty acid esters of glyceryl ethers.

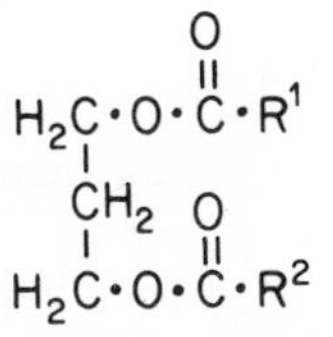

$$H_2C \cdot O \cdot \overset{O}{\underset{\|}{C}} \cdot R^1$$
$$\underset{|}{CH_2} \quad O$$
$$H_2C \cdot O \cdot \overset{O}{\underset{\|}{C}} \cdot R^2$$

Fig. 3.3. 1,3-diacyl propane diol.

In the 19th century, fats were thought to be mixtures of simple (single-acid) glycerides. 100 years of careful analysis and improving techniques has demonstrated their huge complexity.

We can get a good idea of how complex a mixture of glycerides we might expect to find in a tissue by merely considering the possible combinations of three common fatty acids, palmitic (P), oleic (O) and stearic (S). As you can see in Fig. 3.4, eighteen different species are possible, (or 36 when we take into account the possibility of optical enantiomers, discussed on page 104). When one also considers the quite significant quantities of linoleic acid and more highly unsaturated acids

especially in plant glycerides, you can see that the number of possible species is enormous.

Is it so in practice? This calculation assumes that any fatty acid may occupy any of the three positions in relation to any other acid; in other words that the distribution is random. This would mean that during their biosynthesis there would be no selectivity at any step in the pathway. Many people (notably Hilditch, see page 100) considered that this was the case, until some years ago studies with purified preparations of the enzyme pancreatic lipase (which rapidly hydrolyses the outer ester bonds of a triglyceride but not the central one) revealed that the proportion of unsaturated to saturated fatty acids in the resulting 2-monoglyceride was considerably greater than on the outer positions of the original triglyceride. It was argued that as the outer positions were equivalent and could not be distinguished from each other, then the remaining fatty acids must be randomly distributed between these two positions. This was known as the '1,3-random, 2-random' or 'restricted

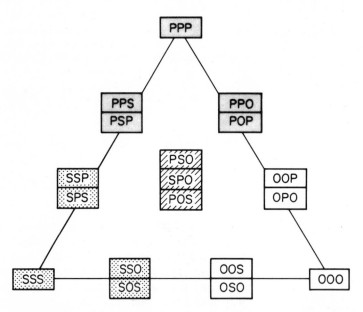

Fig. 3.4. Possible species of triglycerides from three fatty acids.

random' hypothesis. By analysing the composition of total triglyceride fatty acids and either the fatty acids released by lipase hydrolysis or those present in the resulting monoglycerides, and by using the assumptions set forth in this hypothesis, then one can calculate the range of molecular species to be expected in any given fat. The most direct way to find out what species are present in a particular fat is to separate them from each other and analyse each one directly. Only recently has this begun to be possible with the development of TLC on silica gel impregnated with silver nitrate. Details of these techniques are discussed in the final section of this chapter (for

glycerides) and in chapters 1 and 2 (for fatty acids). At the same time, biochemists have also realized that, owing to the inherent asymmetry of glycerol*, the 1- and 3-positions are *not* equivalent. Techniques are now available for distinguishing between positions 1 and 3* and it has become apparent that fatty acids are in many cases not randomly distributed between the outer positions. Complete generalization about glyceride composition is impossible, because while many glycerides are now known to have a *stereospecific* distribution of fatty acids,

* See chapter 1 for an explanation of the numbering system for glycerol derivatives.

others seem to conform to the *restricted random* distribution, while yet others have a completely *random* distribution. Lard is unique in having a preponderance of saturated fatty acids on the 2-position.

The cell has mechanisms for continuously modifying the composition of its complex lipids.

There is an important difference between animals and plants with respect to glyceride composition. Plants must of necessity synthesize their glycerides from simple starting materials according to their requirements. Unlike animals, they also have the ability to synthesize the linoleic acid which they possess in large quantities. The fatty acid composition of animal glycerides, however, is greatly influenced by the diet and therefore, ultimately, by the vegetables they eat. The way in which dietary glycerides are modified by the animal may differ between species and possibly from organ to organ within a species. Such modifications may not only be effected by the animal's own cells. For example, in ruminants such as cows, sheep or goats, the micro-organisms present in the rumen hydrogenate the double bonds of dietary polyunsaturated fatty acids like linolenic and linoleic to form a mixture of mainly saturated and *trans*-monoenoic acids (chapter 2, page 63). An outstanding and significant feature of glyceride fatty acid composition is that it is quite distinctly different from that of either the phospholipids or the non-esterified fatty acid (NEFA) pool. An

understanding of this cannot be obtained simply from analyses of fatty acid distributions but must depend on a study of the different metabolic pathways, and of the individual enzymes of those pathways, involved in the biosynthesis of each lipid class. We shall discuss this in the following section.

There are two pathways for the biosynthesis of glycerides. In the first, triglycerides are synthesized completely from their simplest components; in the second, partial glycerides are re-acylated.

We inferred in the last section that glycerides may arise in different ways depending on whether the tissue needs to start at the beginning from component parts or whether it requires to modify existing — possibly dietary — glycerides. Historically, the *'de novo'* pathway, now usually known as the *glycerol phosphate pathway* was worked out first (Fig. 3.5). This pathway was first proposed by the American biochemist, E.P. Kennedy, based on the earlier work of Kornberg and Pricer, who first studied reaction 2, the formation of phosphatidic acid by acylation of glycerol phosphate. We shall discuss this reaction and the subsequent one — the release of inorganic phosphate by phosphatidate phosphohydrolase — in chapter 4. You may well find it helpful to read the relevant sections of chapter 4, namely 'the biosynthesis of phosphatidic acid' and 'phosphatidate phosphohydrolase' in conjunction with this section, rather than read the book from cover to cover. Indeed, as we pointed out in chapter 1,

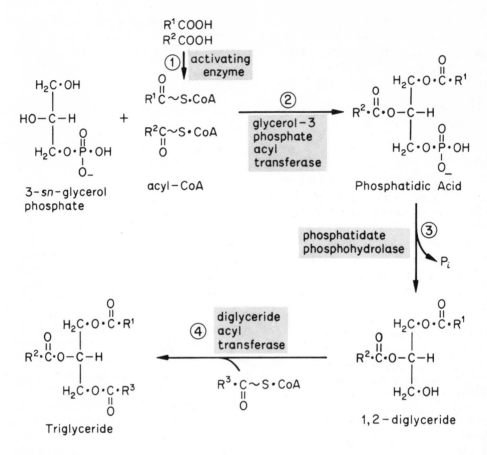

Fig. 3.5. The glycerol phosphate pathway.

this applies to many sections of this book. The acylation steps require the fatty acids to be in the 'activated form' as their CoA thiol esters, a subject which we have already fully discussed in chapter 2. Kennedy also demonstrated the role of phosphatidic acid in phospholipid biosynthesis as well as in glyceride synthesis and one of his outstanding contributions has been to point out that the diglyceride derived from this phosphatidic acid forms the basic building block for both triglycerides and phosphoglycerides.

Cellular concentrations of glycerol phosphate are important in controlling triglyceride biosynthesis

Why does the cell transfer fatty acids to a phosphate ester and then split off the phosphate? Could not the glycerol backbone of the lipid be acylated directly?

In fact, the acyl transferase enzyme is very specific toward 3-*sn*-glycerol phosphate and has no reaction with glycerol itself. The source and availability of glycerol phosphate are therefore very important to glyceride synthesis. Most is derived from glucose by the glycolytic pathway, some *via* the pentose phosphate pathway and a little by direct phosphorylation of glycerol by glycerol kinase. The significance of free glycerol as a precursor for glycerides has not yet been established. It is present in very small quantities in the bloodstream but is metabolized very rapidly. The main factors influencing the amounts of glycerol phosphate available for glyceride synthesis are those which regulate the levels and activity of the enzymes of glycolysis, gluconeogenesis and lipogenesis. These in turn are affected by the nutritional state of the animal. For instance, in rats that have been starved for two days, the level of glycerol phosphate is down to one third of the original concentration but is rapidly restored by refeeding. On the other hand, the levels of the other components which might exert control on glyceride synthesis— the acyl-CoA esters — appear to *increase* during starvation.

The enzymes of glyceride synthesis are located in the endoplasmic reticulum but 'soluble' factors may control the overall reaction.

Another possible point of control has emerged from studies on the subcellular location of the *glyceride synthetase*, as the full sequence of enzymes is called.

Glyceride synthesis occurs in a number of tissues but those most extensively studied have been liver, (where they are assembled as components of lipoproteins for transport to other tissues), intestinal mucosa (where they are synthesized from absorbed components) and mammary gland (which is important in milk fat synthesis). Fractionation of these cells indicates that the microsomal fraction, originating mostly from the endoplasmic reticulum, is the major site of synthesis. However, although the complete synthesis is achieved by microsomal enzymes, addition of the 'supernatant' fraction of the cell stimulates the rate of glyceride formation. When the supernatant fraction is heated, part, but not all, of the stimulating activity is lost. The part which is due to the *heat stable factor* has been identified as a mixture of unsaturated fatty acids. Palmitoyl-CoA has generally been used as a precursor for studies of glyceride synthesis *in vitro* and in view of what we have said about the composition of glycerides, we could interpret this stimulation by unsaturated fatty acids as indicating a preference of the enzyme system to produce mixed acid triglycerides. The *heat labile factor* seems to accelerate the step from phosphatidic acid to diglyceride and may therefore be a 'soluble' form of *phosphatidate phosphohydrolase*. We shall discuss the properties of the different forms of this enzyme in chapter 4. We have stated that acyl-CoA thiolesters are the active species in transferring acyl groups to glycerol phosphate. However, several biochemists have noticed that a mixture of a free fatty acid, coenzyme A, ATP and the supernatant proteins are more effective than acyl-CoA alone in forming phosphatidic acid from glycerol phosphate. In studies with enzyme preparations from the bacteria *E. coli* and *C. butyricum,* which catalyse the acylation of glycerol phosphate, acyl-CoA could be effectively replaced by the fatty acid thiol ester of the acyl carrier protein (ACP). We do not yet know whether the supernatant fraction contains a protein which acts in a similar way to the bacterial ACP and which is responsible for part, at least, of the stimulating effect of this fraction. Studies of the many components of the synthetase and the way in which they are co-ordinated to control glyceride biosynthesis is one of the chief pre-occupations of researchers in this field at present. One of the difficulties we should be aware of is the artificial nature of cell fractionation. We do not know that because an enzyme or component of an enzyme complex is in the *supernatant* fraction after centrifugation of the broken cells, that is is necessarily 'soluble' in the living cell or whether it has been stripped from its original site during the isolation procedure. It could also be that the soluble fraction of the cell has much more 'structure' than has previously been assumed and that 'soluble' components throughout the cytoplasm are closely associated with the various cytoplasmic organelles. Another pitfall for the lipid biochemist

is in the use of acyl-CoA thiol esters; these compounds are strong detergents and when the enzymologist uses them as substrates in enzyme studies *in vitro,* he is always to some extent working under inhibitory conditions. He always has to be aware that the classical Michaelis-Menten kinetics for truly water-soluble substrates cannot necessarily be applied rigorously to compounds which form micelles.

The monoglyceride pathway serves to modify existing glycerides.

The second pathway for glyceride synthesis involves a stepwise acylation of a monoglyceride and is usually called the *monoglyceride pathway* (Fig. 3.6). This reaction, first discovered by G. Hübscher's research team in Birmingham, England, is catalysed by enzymes in the endoplasmic reticulum of the small intestinal mucosa of many species. The 2-monoglyceride isomer is preferred to the 1-monoglyceride and the nature of its fatty acid influences the rate of the first acylation step. Monoglycerides with short chain length saturated or longer chain length unsaturated fatty acids are the best substrates. *Diglyceride acyl transferase* is specific for 1,2-diglycerides and will not acylate the 2,3- or 1,3-isomers. Diunsaturated or mixed acid diglycerides are better substrates than disaturated compounds, but we have to be careful

Fig. 3.6. The monoglyceride pathway for triglyceride synthesis.

when we interpret results of this kind. Lipids containing unsaturated fatty acids are much more easily emulsified than saturated ones, so that we may not be observing differences in specificity of the enzyme but differences in *solubility* of the substrates. The substrate for this pathway, monoglyceride, arises mainly from the hydrolysis of dietary triglycerides by the enzyme lipase in the intestinal lumen. Therefore this mechanism is one by which existing glycerides are modified, rather than one by which new fat is laid down. The processes by which these hydrolysis products are absorbed and transported through the intestinal walls into the epithelial cells where they are resynthesized to triglycerides will be described later in a section on fat absorption. The synthesis of triglycerides can be compared with that of long chain fatty acids in that a sequence of enzymes is involved in handling water-insoluble intermediates. In higher animals, at least, the cell has developed a *multi-enzyme complex*, similar to the fatty acid synthetase, (and which, by analogy, we call *triglyceride synthetase*), in order to deal more effectively with these intermediates. Evidence for this came when the American biochemist, J.M. Johnston, purified *triglyceride synthetase* through several stages and showed that the increase in the specific activity at each stage was the same for all three enzymes in the sequence. Another aspect of the tightly organized structure of the synthetase is that the acyl-CoA is not 'free' but is in some way bound to the protein, suggesting the presence of a molecule analogous to the bacterial ACP.

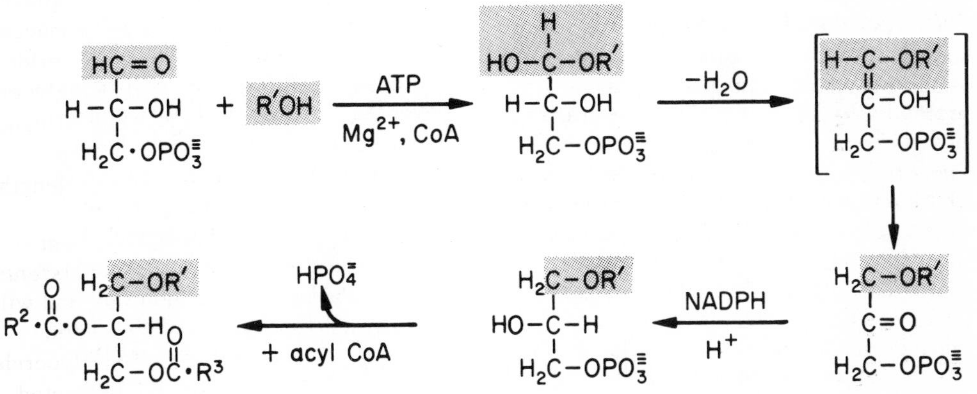

Fig. 3.7. Formation of the ether bond in glyceryl ethers.

*Ether bonds in lipids are
formed by condensation of
a fatty alcohol with the
carbonyl group of
glyceraldehyde-3-phosphate*

There has long been much controversy
but no firm evidence about the mechanism
of formation of ether bonds in glyceryl
ethers. In 1969, F. Snyder and his
colleagues in the U.S.A. proposed a
pathway involving condensation between
a fatty alcohol and the carbonyl group of
glyceraldehyde-3-phosphate. The reactions
are catalysed by enzymes in microsomal
fractions of certain mouse tumours
(Fig. 3.7). The function of ATP and CoA
in this system are not yet understood.

*Lipases hydrolyse preferentially
the outer fatty acids of a
triglyceride*

The existence of enzymes which hydrolyse
the ester bonds of triglycerides has
already been mentioned and is a
prerequisite for the provision of starting
materials for the monoglyceride pathway.
Their role in providing glyceride mixtures
which can easily be absorbed through the
intestinal wall will be discussed in a
later section. There are many enzymes
called *esterases* which hydrolyse ester
bonds in general, but *lipases* form a
distinct class and the distinction rests
on the physical state of the substrate.
The *milieu* in which a lipase acts is
heterogeneous: the lipid substrate is
dispersed as an emulsion in the aqueous
medium and the enzyme acts at the inter-
face. If by some means, a one phase
system is obtained, for example by
employing a short chain length tri-

glyceride such as triacetin or by employing
sufficient quantity of detergent, then the
lipid may be hydrolysed by an esterase
but not a lipase.

Lipases are widespread in nature and
have been found in animals, higher plants
and micro-organisms. The initial step
in the hydrolysis is the splitting of the
fatty acids esterified to the primary
hydroxyls of glycerol. Even though these
are now known to be distinguishable,
this reaction is not stereospecific and
the fatty acids in the 1- and 3-positions
are initially removed at equal rates. Once
one fatty acid has been removed, however,
the resulting diglycerides and subsequently
the monoglycerides are more slowly
hydrolysed than the original triacylglycerols.
The preference for removal of the 1 and 3
fatty acids together with the diminished
rate of hydrolysis of partial glycerides
results in an accumulation of monoglycerides as the primary products of
lipase hydrolysis. This is especially
the case *in vivo* where controls such as
resynthesis of di- and triglycerides,
regulate the degree of hydrolysis; *in vitro*
the hydrolysis is much more likely to go
to completion, especially if a high concentration of bile salts is added as a
detergent to effect better emulsification
of the substrate. Lipase not only
hydrolyses fatty acids on the outer
position of glycerides but will also
liberate the fatty acid esterified in the
1-position of phosphoglycerides. In most
cases, the rate of hydrolysis is independent of the nature of the fatty acids
released. However there are a steadily
growing number of exceptions to this
rule: fatty acids with a chain length less

than twelve carbons, especially the very short chain acids of milk fats, are cleaved more rapidly than normal chain length acids, while the very long polyenoic acids, $C_{20:5}$ and $C_{22:6}$ found in the oils of fish and marine mammals are hardly hydrolysed at all. Recently, too, a lipase has been discovered in the microorganism *Geotrichum candidum* which seems to be specific for oleic acid in whichever position it is esterified.

Most of the studies on lipase which we have described here have been done with the enzyme isolated from pancreas or pancreatic juice; this is the enzyme which is important in the absorption of dietary glycerides. Other kinds of lipases are present in other tissues and differ mainly in their substrate specificities. One of the most important of these is *lipoprotein lipase* sometimes known as *clearing factor lipase* (for a further discussion see chapter 6). This enzyme is distinguished by the fact that it hydrolyses chiefly triglycerides associated with proteins such as chylomicrons and serum lipoproteins. The enzyme will hydrolyse emulsions of simple triglycerides *in vitro* much more rapidly if a protein is added to the incubation mixture. A recent description of the purification of lipoprotein lipase illustrates beautifully the way in which the specific property of an enzyme to form a complex with its substrate can be used in enzyme purification. C.J. Fielding added lipoproteins to his partially purified enzyme: this created a complex with a density lower than that of the original enzyme and of much of the contaminating protein. When centrifuged

in a medium of suitable density, the complex 'floated' to the top of the centrifuge tube and could be separated from the more dense material.

Another lipase, present in adipose tissues, is activated by hormones such as adrenalin. Many of its properties distinguish it from pancreatic and lipoprotein lipases but little is known about how it fits into the overall controlling processes of lipid metabolism. Adipose tissue also contains a lipase which is much more active on monoglycerides than are other lipases. Enzymes with lipase activity are still being discovered and it will be a long time before the whole complex jigsaw of glyceride breakdown and resynthesis can be pieced together.

As in all branches of lipid biochemistry, better separation methods have given impetus to metabolic studies.

During the last century, most chemists (with the notable exception of Berthelot) assumed that natural fats were mixtures of simple, single acid glycerides. The lack of techniques for separating fat components prolonged this state of ignorance. A complete glyceride analysis resolves itself into two parts. First, a separation of the glycerides themselves into components; then a fatty acid analysis of the different components. The chief methods for separating glyceride components are in order of historical importance: fractional crystallization; low pressure fractional distillation; counter-current distribution; argentation TLC and gas-liquid chromatography. The epic work which has provided the solid basis of our knowledge was that of Hilditch and his school beginning in 1927. They refined the technique of fractional crystallization and also introduced an oxidation method for the determination of trisaturated glyceride species. This depends on the fact that all glycerides containing unsaturated fatty acids could be oxidized by permanganate in acetone to yield oxidized fats whose physical properties were very different from the saturated fats and which could therefore be easily separated from them. It was these techniques that led Hilditch to develop his theories of *even* or *random* distribution. The next step forward came with the introduction of counter-current distribution, a technique which separates mixtures according to their partition between immiscible solvents. The invention of an automatic apparatus which could perform a large number of transfers enabled a much higher degree of resolution of natural fats, especially highly unsaturated ones, than that achieved by crystallization. Nevertheless the method is quite laborious, consumes large amounts of solvents, and is not now widely used for small scale analytical work.

Modern methods of analysis use a combination of argentation TLC, lipase hydrolysis and gas-liquid chromatography.

Two methods in combination have been responsible for most of the present day analyses of natural fats — lipase hydrolysis and thin layer chromatography — each combined with gas-liquid chromatography. The basis for the lipase method has been mentioned earlier in the chapter. It depends on the fact that (1) the hydrolysis yields mainly monoglyceride and (2) the outer fatty acids only are removed so that the monoglyceride in question is the 2-isomer (Fig. 3.8). In fact, in the laboratory, it is quite easy to take the hydrolysis to completion, which would distort the results, so that usually the hydrolysis is stopped when 65% of the maximum yield of mono-glyceride has been formed. The partial

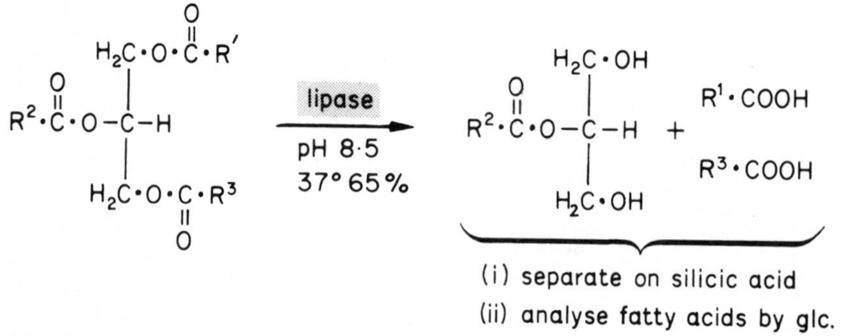

Fig. 3.8. Application of lipase to analysis of glycerides.

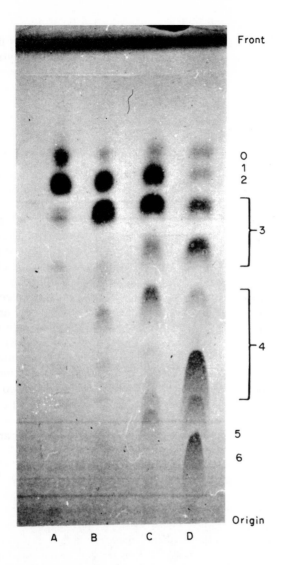

Fig. 3.9. Thin layer chromatography of natural triglyceride mixtures on silica gel G impregnated with silver nitrate (5% w/w). Solvent: isopropanol–chloroform, 1·5:98·5 (v/v). Spots were located by spraying with 50% sulphuric acid and charring.

A. palm oil.
B. olive oil.
C. groundnut oil.
D. cottonseed oil.

The numbers represent the total number of double bonds in each triglyceride molecule.

glycerides and free fatty acids can be separated on columns of alumina or silica, the glyceride fractions saponified to release their fatty acids which can then be analysed as their methyl esters by GLC. The calculation of the proportions of components in the original mixture depends on the assumption that the acids released by lipase were originally randomly distributed between the 1 and 3-positions. This may be the case for some fats but certainly not for all.

The most up-to-date method, and the one most capable of the best resolution is argentation TLC (Fig. 3.9), in fact it is the only method whereby a unique separation of a glyceride mixture has been achieved. Even so, it is dependent on the total number of double bonds in the molecule, so that in a complex mixture one must expect there to be several components which cannot be resolved. Identification of the components depends on being able to synthesize model glycerides; this may impose a practical limitation on the method. However, by fractionating the glycerides into groups by argentation TLC and then subjecting each group to enzymic hydrolysis, a complete analysis of the mixture can usually be made. Methods have been introduced for the fractionation of whole glycerides (as opposed to their fatty acid constituents) by GLC but it is doubtful whether the resolution will ever be as great as that achieved by argentation TLC.

The realization that the 1- and 3-positions of glycerol are not equivalent led several biochemists to try to find methods for distinguishing between them and hence to discover whether there were differences in the fatty acids occupying those positions. This is known as

stereospecific analysis. One of the first and most frequently used methods for stereospecifically analysing all three positions of a triglyceride was devised by the Canadian biochemist, Hans Brockerhof. His scheme is depicted in Fig. 3.10a. It depends on the facts that (a) lipase non-specifically cleaves the fatty acids in the 1- and 3-positions yielding a 2-monoglyceride, (Step 1) and (b) phospholipase A *specifically* cleaves the fatty acid in position 2 of a phosphoglyceride (step 3; see also chapter 4). First the total fatty acid composition of the original triclyceride is obtained by saponification and GLC. Then the 2-position fatty acid is obtained by analysis of the monoglyceride and the 1-position fatty acid by analysis of the lysophosphatidyl phenol. The 3-position fatty acid is obtained by difference. The disadvantages of this method are, first, that lipase hydrolysis does not yield a random sample of fatty acids when the triglyceride contains (a) long chain polyunsaturated fatty acids such as $C_{20:5}$ and $C_{22:6}$ which are hardly hydrolysed at all, or (b) very short chain fatty acids which are hydrolysed much more rapidly than normal ones. Secondly, the calculation of the 3-position fatty acids by difference is inaccurate for minor components. To eradicate these difficulties, Brockerhof devised an alternative method which eliminates the use of lipase (Fig. 3.10c). A third method has been invented by W.E.M. Lands at the University of Michigan. This beautifully illustrates the asymmetry of glycerol by the use of an enzyme, *diglyceride kinase*, which phosphorylates *only* the 3-hydroxyl of a 1,2;diglyceride and *not* the 1-hydroxyl of a 2,3-diglyceride (Fig. 3.10b)

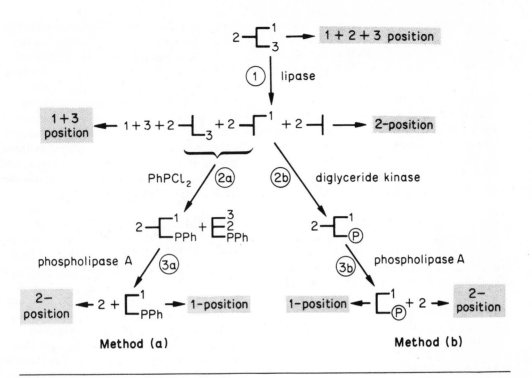

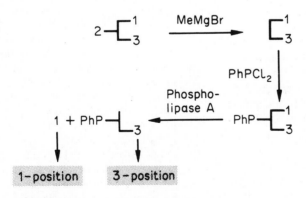

Method (c)

Fig. 3.10. Stereospecific analysis of triglycerides.

As these methods have increased in use, so more and more natural glycerides have been shown to possess a *stereospecific* distribution of fatty acids rather than a *random* or *1,3-random* one. Examples of this are the milk fats, in which the characteristic short chain fatty acids accumulate in position 3. Animal depot fats have saturated fatty acids at position 1, short chain and unsaturated acids at 2; position 3 seems to have a more random population although polyenoic acids tend to concentrate at position 3 in mammals but in position 2 in fish and invertebrates. Certain seed oils contain acetate residues which only occur at the position 3. This inherent assymmetry may lead to optical activity, which although extremely small, is measurable especially in the extreme cases such as the glycerides containing acetate.

Triglycerides are efficient reservoirs of energy in animals and some plants.

We have discussed the enormous complexities of glyceride composition, their formation and breakdown. What then is their purpose? Glycerides present the only examples of tissue lipids whose function is quite unequivocal. They represent a very convenient and highly efficient form of storing fatty acids. Fatty acids constitute a major reserve source of fuel in animals and oil bearing plants. During the controlled oxidation of fatty acids, ATP is generated yielding an energy supply equivalent to 9 kcal/g of fatty acid catabolized, compared with 4 kcal/g of carbohydrate or protein. Carbohydrate, moreover, has to be stored in a bulky hydrated form, whereas three fatty acids can be stored per triglyceride molecule in anhydrous form in adipose tissue. The real importance, therefore, of triglycerides, is that they enable the animal or plant to store a much larger reservoir of energy than would be possible by storing carbohydrate or protein.

Mobilization of fat is controlled by nutritional and hormonal factors

When triglyceride, stored in the adipose tissue, is required as an energy source (e.g. when an animal is fasting or doing muscular exercise) it must be *mobilized* and transported to other tissues which need fuel. The mobilized form of fat is non-esterified fatty acid which is released from the triglyceride by the combined action of a *hormone sensitive lipase* and other lipases (see p. 99). The fatty acids released by these lipases may follow alternative pathways depending on the nutritional and metabolic state of the animal. If a plentiful supply of glycerol-3-phosphate is available, the fatty acids may be re-esterified to form triglycerides by the pathway we have already described. If no glycerol-3-phosphate is available, the fatty acids diffuse into the blood and are transported to other tissues as the albumin complex (see Chapter 6] and are there oxidized by β—oxidation (see Chapter 2) to provide energy or re-esterified. The other product of lipase hydrolysis in the adipose tissue is glycerol. In other tissues this would be converted into glycerol-3-phosphate by *glycerol kinase*. This enzyme, however, does not occur in adipose tissue and therefore the only source of glycerol-3-phosphate is carbohydrate catabolism. Thus, in periods of starvation, when little glucose is being metabolized, there is a high rate of fat mobilization. The reverse is true in fed animals.

Other factors, in addition to the availability and utilization of carbohydrate, influence the rate of fat mobilization from adipose tissue. Hormones such as adrenalin promote fat mobilization by activating hormone sensitive lipase, which appears to be the rate limiting enzyme in adipose tissue lipolysis. This activation is mediated *via* cyclic-AMP by an as yet unknown mechanism. The activation may be antagonized by prostaglandins (see Chapter 2) thus implicating the prostaglandins in the control of fat mobilization, although it should be emphasized that we do not know whether prostaglandins are *physiologically* important in this respect. Insulin strongly inhibits fat mobilization, first by promoting glucose metabolism and increasing the availability of glycerol-3-phosphate, and also by inhibiting hormone sensitive lipase by lowering the concentration of cyclic-AMP in adipose tissue. Insulin secretion is itself stimulated by an increased plasma concentration of glucose and thus we can see that nutritional and hormonal factors are inseparable and interact to exert a fine control on lipid metabolism. A much more detailed discussion of the physiological chemistry of fat mobilization and lipid-carbohydrate relationships can be found in reference 5 at the end of this chapter.

The Absorption of Fats

(i) Non-ruminants

Fat in foods consists largely of triglyceride with a small proportion of phospholipid and in the case of processed foods, of surface active additives such as monoglycerides. In physical form it may vary from discrete particles to dispersions or even lipoproteins and the digestive system has to be capable of handling all these forms. After ingestion, the first process, carried out in the stomach, is the formation of an oil-in-water emulsion (probably stabilised by phospholipid) due to the mechanical movements of churning, etc. produced by gastric mobility. Lipoproteins are broken down by proteolysis liberating the lipids but little lipolysis or absorption takes place in the stomach; this happens only after the contents are passed in small amounts into the duodenum.

Secretion of pancreatic juice and bile into the duodenum initiates the absorbtive process by virtue of (a) the lipolytic action of pancreatic lipase and phospholipase A and (b) the marked ability of micelles of the conjugated bile acids to solubilize partial glycerides and fatty acids.

The initiating process is attack by pancreatic lipase on triglycerides at the oil/water interface of the fat particles. The reaction:

$$\text{Triglyceride} \xrightarrow{\text{fast}} \text{free fatty acid} + 1{:}2\text{-diglycerides} \longrightarrow 2\text{-monoglycerides} + \text{free fatty acids}$$

$$\xrightarrow{\text{slow}} \text{free glycerol} + \text{free fatty acid}$$

generates two types of surface active agent, fatty acids and monoglyceride. Any phospholipid present is degraded by phospholipase A (see chapter 4) to lysophospholipid, also a powerful detergent. However the major components in the duodenum are bile salts, free fatty acids and 1,2-diglycerides. The bile salts, themselves products of breakdown of cholesterol in the liver, are principally the glycine and taurine conjugates of tri- and di-hydroxycholanic acids and are anionic detergents that readily form mixed micelles with fatty acids and monoglycerides. Thus in the early stages of digestion there will exist a whole range of particles ranging from coarse emulsion particles of 10,000 Å in diameter that are progressively decreased in size due to lipolytic attack, down to micelles of pure bile salts of 40 A diameter.

When held in these mixed micelles, long chain 1-monoglycerides are more susceptible to attack by pancreatic lipase than are the 2-monoglycerides or even short chain 1-monoglycerides. The species primarily absorbed by the brush border (Fig. 3.11 and 3.13) are 2-mono-glycerides and free fatty acids. The bile salts themselves are not absorbed in the proximal small intestine but pass on to the ileum when absorption occurs and

they are recirculated in the portal blood to the liver and then to the bile for re-entry at the duodenum. The general processes involved in the first stages of fat dispersion, hydrolysis and absorption are indicated schematically in Fig. 3.12.

Thus at the border of the intestinal epithelial cells (Figure 3.11 and 3.13) the mixed micelle is broken down and free fatty acid and monoglyceride enter the cells. The mechanism of entry is still unclear. Pinocytotic activity appears insufficient to account for the speed of absorbtion and one is left with molecular or 'soluble' absorption. Certainly during their passage through the intestinal epithelial cells, fatty acids are converted via their acyl-S-CoA esters into triglycerides. This change from water dispersible derivatives to compounds of zero solubility would give rise eventually by coalescence to the stabilised fat droplets visible in electron micrographs of actively absorbing gut. The stabilisation process involves the endoplasmic reticulum, especially the Golgi apparatus, and it may well be here that the protein components (as well as phospholipid and cholesterol) necessary for particle stabilisation are united to the triglyceride to form chylomicrons. The chylomicrons pass from the cells to the intercellular spaces by reverse pino-cytosis, then to the lacteals and to the lymphatic channels and are distributed

Fig. 3.11. Light microscopic picture of the microvilli (brush border) of pig gut. Magnification × 70.

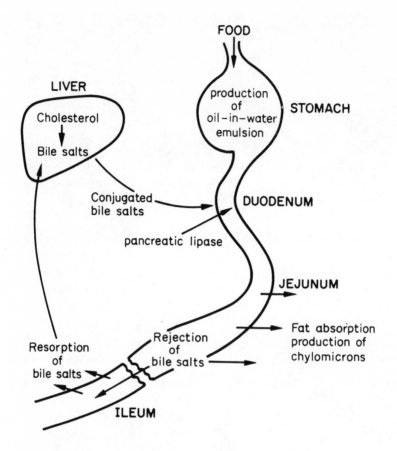

Fig. 3.12. Schematic diagram of the processes involved in the first stages of hydrolysis and absorption.

throughout the body by the blood stream. The chylomicrons, irrespective of the nature of the fat fed, consist mainly of long chain triglycerides (above 14C atoms) since the short chain acids (presumably because of their greater water solubility when liberated in the lumen) are selectively transported into the mesenteric portal blood as free fatty acids.

The general processes occurring during absorption are indicated schematically

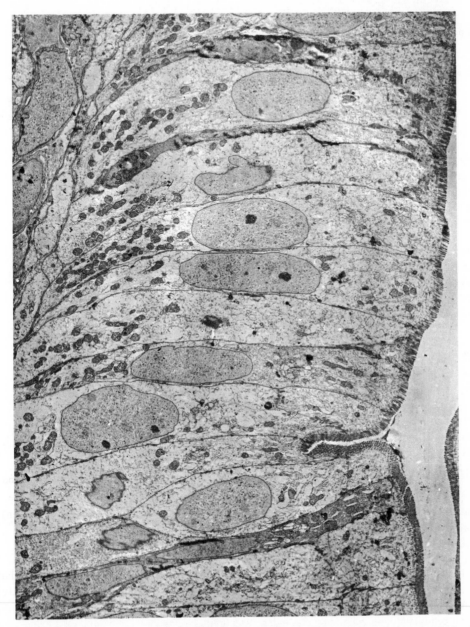

Fig. 3.13. Electron micrograph (× 2600) of epithelial cells of pig intestine showing the length of the cell relative to the brush border, and the large number of mitochondria.

in Fig. 3.14 so far as our knowledge goes.

Malabsorption Syndromes

Idiopathic Steatorrhea and Sprue. These conditions present an inability to absorb fat and hence show massive excretion in the faeces, coupled often with diarrhea, the patients being thus perforce on a very low fat diet. The causation varies from bacterial invasion of the gut to sensitisation by gluten in the diet. The profound changes in bacterial population of the gut, both in level and composition, make it difficult to decide whether these changes are primary or secondary events. The excreted fat is derived not only from unabsorbed dietary material but also from the organisms themselves and from tissue breakdown.

The bacteria undoubtedly affect the composition of the excreted fat. For example a major component (absent from the diet) of faecal fat, 10-hydroxystearic acid was shown by tracer studies to be formed by bacteria from stearic acid.

Because of the low fat absorption patients with I.S. are as close to deficiency in essential fatty acids as any group yet studied. The linoleic acid of the plasma is often less than 50% of normal, and is replaced by oleic and palmitoleic acid. Surprisingly however the level of arachidonic acid is normal even though this must have been derived largely from dietary linoleic acid. This would suggest some sparing effect on the linoleic acid and effective diversal away from oxidation.

Essential Hyperlipaemia

This name covers a group of conditions all of which show elevated plasma triglyceride levels i.e. above 400 to 800 mg/100 ml. There may be no marked symptoms, but frequently xanthomatosis and less frequently liver and spleen enlargement may be found. The conditions are usually induced by fat or carbohydrate in the diet. Treatment consist of reducing fat intake below 13% of total calories in the former case or carbohydrate intake below 25% in the latter.

(ii) Ruminants

The lipids of ruminant animals differ from those of non-ruminants in two respects: firstly the high levels of stearic acid in some depot fats and secondly the occurrence of branched chain fatty acids in tissue and milk lipids. The processes occurring in the more complex alimentary tract system of the ruminant can be summarized as follows.

(a) In the rumen a complex population of micro-organisms split all types of dietary lipid to release free fatty acids. The unsaturated acids undergo isomerization and biohydrogenation (see page 63, chapter 2) to produce saturated and *trans* positional isomers. Little or no absorbtion takes place here. The fermentive activities give rise to large amounts of acetic and propionic acids and all these materials pass through the true stomach to the small intestine together with branched chain and odd numbered saturated fatty acids derived from the rumen bacteria.

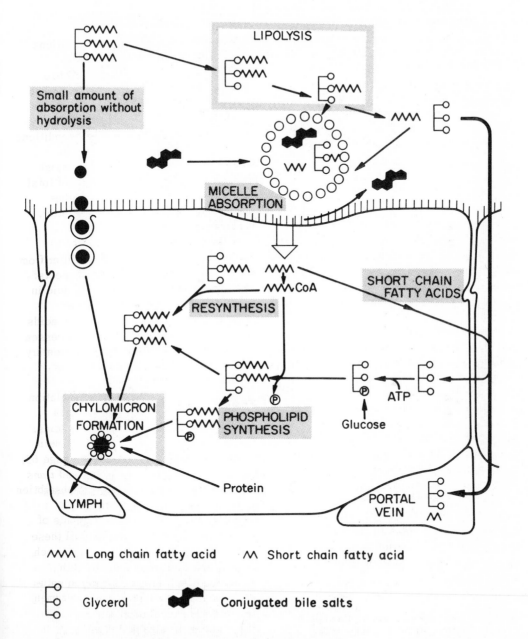

LIPOLYSIS

Small amount of absorption without hydrolysis

MICELLE ABSORPTION

SHORT CHAIN FATTY ACIDS

RESYNTHESIS

CoA

ATP

Glucose

CHYLOMICRON FORMATION

PHOSPHOLIPID SYNTHESIS

LYMPH

Protein

PORTAL VEIN

〰〰〰 Long chain fatty acid   〰 Short chain fatty acid

Glycerol   Conjugated bile salts

Fig. 3.14. Schematic diagram of digestion and absorption of fat.

(b) During passage through the small intestine a micellar solution of free fatty acid and conjugated bile acids (possibly with some lysophosphatidylcholine derived from biliary phosphatidyl choline) is formed and absorption of the fatty acids occurs. Unlike the non-ruminant there is little or no glycerol or monoglyceride of dietary origin to be absorbed because they are removed by fermentation and hydrolysis in the rumen.

(c) The triglycerides formed in the intestinal epithelial cells are derived largely from the absorbed saturated acids and *trans* unsaturated acids accounting for the difference in composition of the depot fats from non-ruminants.

STEROL ESTERS

For the purposes of this book we have defined 'lipids' as esters of long chain fatty acids. This eliminates the large and metabolically important group of water-insoluble compounds, the steroids.* In a great many organisms and tissues, sterols exist as mixtures of the free alcohols and their long chain fatty esters and it is the latter compounds which come within the scope of this book. Trying to simplify in this fashion leads us into the difficulty that the distinction between sterols and their esters is very artificial and it is almost impossible to talk about the esters without mentioning the parent compounds. Nevertheless we shall not deal with the biosynthesis of the steroid ring system, nor discuss the metabolism of steroids and their derivatives such as the steroid hormones or bile salts, but confine our discussion to the enzymes involved in the formation and hydrolysis of sterol* esters, and their metabolic importance.

Although sterols are widespread in nature, far more work has been done on mammalian sterol metabolism than on that of other members of the animal kingdom or of plants. Also, because cholesterol is the major sterol of mammals, most of our knowledge is of cholesterol ester metabolism and, to the biochemist who deals with animals at least, *cholesterol ester* is almost synonymous with *sterol ester*.

The proportion of free sterol to sterol ester varies from tissue to tissue, species to species.

Cholesterol was discovered as a major component of gallstones in the 18th century. The French chemist Chevreul partly characterized it in 1816 and called it *cholesterine*. Later it was shown to be present in alcoholic extracts of blood and in 1859 Berthelot identified it as an alcohol and prepared cholesterol esters by heating the sterol with fatty acids at 200°. In 1895, cholesterol palmitate and stearate were crystallized from extracts of the serum of dogs and other animals. The structure of cholesterol esters is

* The word *steroid* denotes any substance which is a derivative of the condensed ring system cyclopentanoperhydrophenanthrene. *Sterols* are those members of the steroid class which contain a hydroxyl group capable of forming an ester.

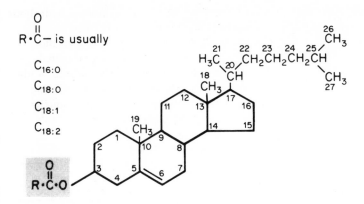

Fig. 3.15. Cholesterol ester.

shown in Fig. 3.15.

Although sterols are present in most mammalian body tissues, the proportion of sterol ester to free sterol varies markedly. For example, blood plasma, especially that of man, is rich in sterols and like most plasma lipids they are almost entirely found as components of the lipoproteins; about 60–80% of this sterol is esterified. In the adrenals, too, where cholesterol is an important precursor of the steroid hormones, over 80% of the sterol is esterified. In brain and other nervous tissue, where cholesterol is a major component of myelin, virtually no cholesterol esters are present.

The major sterol in higher plants is β–sitosterol. Chemists used to think of cholesterol as the 'animal sterol' and β–sitosterol as the 'plant sterol', but now cholesterol has been discovered in many higher plants and algae and is especially abundant in the red algae. Not very many quantitative data are available about sterols in plants, but some at least are present as sterol esters. Much of

the sterol in plants is combined with sugars as sterol glycosides and in some cases the sugar moiety may be acylated with fatty acids. These compounds are known as *esterified sterol glycosides*.

Bacteria have always been regarded as unique in having no sterols. Until recently, *green* algae were thought to be the lowest form of plant life in which sterols occurred: but then they were discovered in the blue-green algae (more primitive than green algae with many features resembling bacteria) and also in the bacterium *Escherichia coli*. It appears that these sterols are rather tightly bound and are not extracted with the usual lipid solvents, chloroform-methanol or ether-alcohol. Only after saponification with alkali followed by solvent extraction of the *completely broken cells* can this tightly bound sterol be removed. Of course, after such treatment we cannot be sure whether the sterol was originally present in the free or esterified form!

Only free cholesterol can be absorbed by the intestine. Subsequently, re-esterification occurs before the reformed esters are transported in the lymph.

Before dietary cholesterol esters can be absorbed by the intestine, they must first be hydrolysed to free cholesterol and fatty acids. Absorption is aided by the hydrolysis products of the other fats, such as monoglycerides, fatty acids and also by phospholipids and bile salts. Experiments *in vitro*, however, using small segments of intestine turned inside out (known as 'everted sacs') apparently demonstrate direct uptake of cholesterol esters without hydrolysis and without the requirement for bile salts. This is a good example of a case where experiments *in vitro* do not provide physiologically valid results and underline the necessity to treat such data critically.

Most cholesterol reaching the mucosal cells is re-esterified, incorporated into lipoprotein particles called *chylomicrons* (see chapter 6) and passed out into the lymph in this form. Chylomicron cholesterol esters are removed, mainly by the liver (but to a small extent by other tissues) and slowly hydrolysed. There may be resynthesis in the liver, possibly with change in fatty acid composition depending on the specificity of the liver enzymes. The resynthesized esters are released from the liver into the plasma as components of lipoproteins and they may be yet again modified by specific plasma enzymes. Further redistribution to different tissues may take place *via* the plasma lipoproteins.

These processes of hydrolysis, absorption, resynthesis, release and uptake by different tissues require the participation of enzymes capable of catalysing both the hydrolysis and synthesis of cholesterol esters. There are three main types of enzymes known to be involved and in the rest of this discussion we shall describe the properties of each in turn.

Sterol ester hydrolase: E.C. 3.1.1.13. Sometimes called cholesterol esterase. Catalyses a low energy reaction not requiring ATP or Coenzyme A.

The first type of enzyme catalyses the reversible reaction:

Cholesterol + non-esterified fatty acid

$\xrightarrow[\text{enzyme-bile salt complex}]{\hspace{2cm}}$ cholesterol ester.

Confusion has arisen over the old name *cholesterol esterase* which implies a single enzyme responsible only for the hydrolysis of cholesterol esters. Even though the enzyme from pancreas has been purified four hundred-fold, during which time the ratio of esterification to hydrolytic activities remained constant, it is still not certain whether these activities are due to a single enzyme, for the preparation contained at least two proteins. There are other pointers to the presence of two distinct enzymes; for example hydrolysis has a higher pH optimum than esterification. We need to interpret these data cautiously, for the pH optimum may be affected by the form of the substrate; cholesterol oleate added in alcoholic dispersion is hydrolysed most effectively at pH 8·6, but in an albumin-bile salt emulsion the best pH is 6·9. The role of bile salts is not completely understood. They certainly act as detergents in aiding emulsification of the substrate and they protect the enzyme against tryptic hydrolysis: but they have a much more specific role, since only the trihydroxy bile salts, such as cholic acid, are effective. Another problem to be solved is: where does the energy for the reaction come from? The synthetic process does not require ATP or CoA and seems therefore to be a low energy reaction. More studies on the nature of the enzyme-substrate complex and the active site should help in solving these mysteries.

Acyl–CoA: cholesterol-O-acyl transferase. The enzyme found in liver and adrenal gland requires ATP and Coenzyme A.

The cholesterol esterifying activity found in liver used to be thought to be due to sterol ester hydrolase working in reverse, but in 1958, S. Mukerjee proved that ATP and CoA were necessary for this reaction and that acyl CoA would effectively replace fatty acid, CoA and ATP:

Acyl–CoA + cholesterol ⟶ cholesterol ester.

This enzyme is located almost entirely in the subcellular organelles, the mitochondria and the endoplasmic reticulum and its function is the reesterification of cholesterol as we described earlier in the chapter. The resynthesized esters

are built into lipoproteins in the liver and later put out into the plasma.

1,2-diacyl glyceryl phosphoryl choline (lecithin): cholesterol-O-acyl transferase. A plasma enzyme which modifies and aids in redistribution of plasma lipoprotein cholesterol esters.

The French workers E. LeBreton and J. Pantaléon and later J. Etienne and J. Polonowski first noticed a connexion between phospholipid hydrolysis and cholesterol ester formation in the plasma; they suggested that fatty acids released from the phospholipid, phosphatidyl choline by phospholipases were subsequently incorporated by means of another enzyme into cholesterol esters. The American biochemist, J.A. Glomset, was the first to show that the reaction was due to a single acyl transferase which could catalyse the transfer of fatty acids (in this case labelled specifically with ^{14}C) from linoleoyl phosphatidyl choline,* but not from albumin-bound linoleic acid, to cholesterol:

Further pieces of evidence for this mechanism were that, (a) the molar increase of cholesterol ester was identical to the molar decrements of free cholesterol and phosphatidyl choline, and (b) the fatty acids which newly accumulated in cholesterol esters during the reaction (mainly polyunsaturated acids) resembled the former pattern of fatty acids in *position two* of phosphatidyl choline. The equation in Fig. 3.16 cannot quite

* For details of phospholipid naming and structure see chapter 4.

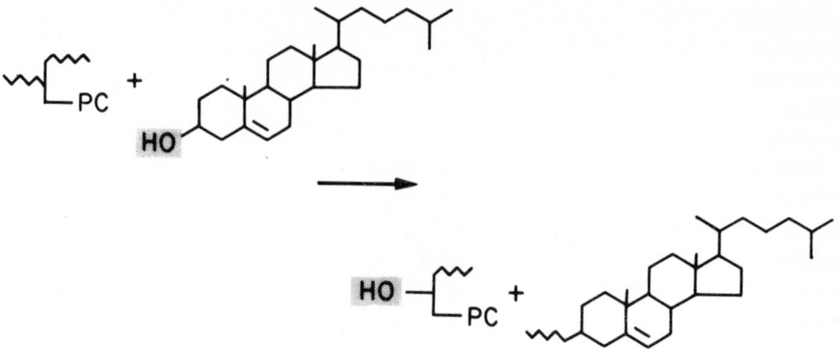

Fig. 3.16. Plasma lecithin: cholesterol transacylase reaction.

explain all the observations. For instance, more saturated fatty acids are transferred to cholesterol than are released from phosphatidyl choline and the amount of lysophospholipid formed is much less than expected. It is possible that some fatty acids are transferred from the *1-position* of the phospholipid or that triglyceride fatty acids can participate in the reaction. Another discrepancy is that in experiments *in vitro*, the enzyme forms cholesteryl arachidonate less readily than would be expected from the amount of this ester present in plasma, whereas cholesteryl linoleate is produced at the expected rate. This may indicate either that the isolated enzyme distinguishes differently between the two fatty acids or that there are different phosphatidyl choline *species* containing each of these fatty acids and occupying different sites on a lipoprotein or different lipoproteins. Certainly the lipoproteins from which phosphatidyl choline *disappears* are different from both those on which cholesterol esters *accumulate* and to which lysophosphatidyl choline is transferred. The fact that the products

are removed to different lipoprotein sites may be one of the factors helping to drive the reaction in the direction of cholesterol ester synthesis.

Questions which have not been complete resolved are whether the liver is the source of plasma *lecithin: cholesterol acyl transferase* as some evidence suggests and whether the composition of plasma cholesterol esters is mainly controlled by this enzyme or whether the essential features of their composition is determined at the stage of liver lipoprotein synthesis (see chapter 6). The physiological role of *lecithin: cholesterol acyl transferase,* and of cholesterol esters themselves is still largely a matter of speculation. Some people have suggested that cholesterol esters may be a transport form of cholesterol or essential fatty acids. This is unlikely in view of the relative slowness with which cholesterol esters are exchanged between one lipoprotein and another, compared with the ready exchangeability of cholesterol. The possibility must be considered that the enzyme may be more important in phospholipid metabolism than in cholesterol metabolism. One of the products, lyso-phosphatidyl choline is a powerful detergent and likely to have rather drastic effects on cells if the concentration is not strictly controlled.

Most of the cholesterol which accumulates in the aorta during atherosclerosis — one of the most menacing diseases of modern times — is in the esterified form. In view of the importance of cholesterol ester metabolism in heart disease, a solution to some of these problems is of the utmost importance. We shall end this section with a brief description of atherosclerosis and its connection with cholesterol ester metabolism.

Ischaemic Heart Disease (I.H.D.) i.e. reduced blood supply to the heart, may result, in man, from accumulation of lipids on the arterial wall. In other animals such accumulations do not often lead to I.H.D.

Although there is still no definitive evidence to show that lipids, or abnormal lipid metabolism, are causative agents in the development of coronary artery disease, there is no question that lipid levels in tissues are affected. Animal experiments with a number of species have shown that increased levels of both saturated fats and cholesterol in the diet give rise to elevated plasma cholesterol levels and to *atherosclerosis*. Atherosclerosis itself consists of localised abnormalities of the inner wall of the artery, usually consisting of accumulation of lipids (particularly cholesterol esters, phospholipid and cholesterol) together with mucopolysaccharides etc. The earliest signs of

the condition are small areas, stainable with lipid stains, that apparently develop further by accumulating lipid, and by proliferation of connective tissue to give degenerative lesions. The artery wall is locally thickened and loses elasticity. The surface presented to the blood stream is thus biochemically and mechanically altered.

The fatal episode in Ischaemic heart disease is the formation of a *thrombus*, or blood clot in a coronary artery, usually at the site of an atherosclerotic lesion, so that the blood supply to the heart is cut off. This failure of blood supply (or *myocardial infarction*) leads either to degeneration of part of the contractile heart tissue and then replacement by non-contractable scar tissue, or else to complete cessation of heart beat. No experimental animal has yet been found in which this fatal episode can be reproducibly produced and for this reason experimental study of this disease has always been difficult. White cells are particularly active in clot formation and recently, evidence has been forthcoming that the prostaglandins (see chapter 2) are involved in clotting and provide a connection with the polyunsaturated acids of the diet.

There is much argument as to whether dietary modification by increasing the level of linoleic acid and decreasing the level of saturated acids, is necessary or not. It is certainly true that plasma cholesterol levels can be lowered by such dietary changes and patients with lipidoses which lead to marked elevation of of plasma lipid (see chapter 6) do have an increased incidence of Ischaemic heart

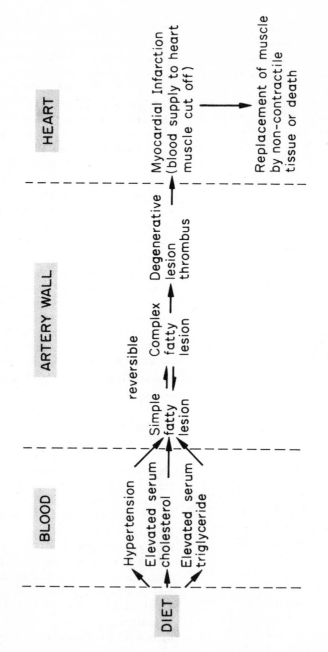

Fig. 3.17. Possible processes involved in the development of ischaemic heart disease.

disease. Studies, albeit on a limited scale, now under way should eventually show whether this action decreases the incidence of the disease. Patients who have already had one infarct are less likely to respond because of the irreversibility of the advanced atherosclerotic lesion.

VITAMIN A ESTERS (RETINYL ESTERS)

The same remarks will apply to vitamin A and its esters as we made in the case of sterols. We shall not discuss the metabolism of vitamin A itself, nor say very much about its role in the biochemistry of vision but confine ourselves to the formation and hydrolysis of its long chain fatty acid esters.

Vitamin A is important for vision, growth and reproduction

Vitamin A is an isoprenoid alcohol, and to avoid confusion with the multiplicity of derivatives and metabolites that arise naturally by oxidation, reduction and isomerization, the parent compound is best called more specifically, *all-trans-retinol* (Fig. 3.18).

The vitamin is important in animal growth, reproduction and vision, but only in the latter case can its role be adequately defined at the molecular level. The enzymic oxidation product of retinol, 11-*cis*-retinal (Fig. 3.18) complexes with a protein, opsin, in the retinal rods to form the conjugated protein rhodopsin. The visual process consists of a series of reactions triggered by the photochemical isomerization of 11-*cis*-retinal to all-*trans*-retinal.

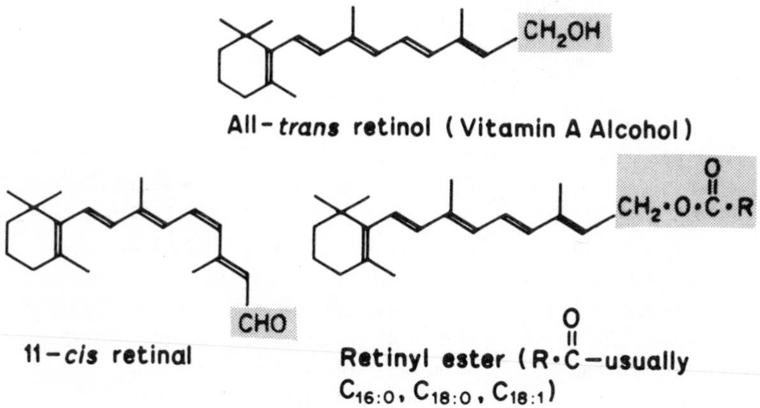

All-*trans* retinol (Vitamin A Alcohol)

11-*cis* retinal

Retinyl ester (R·C—usually $C_{16:0}$, $C_{18:0}$, $C_{18:1}$)

Fig. 3.18. Vitamin A derivatives.

Vitamin A esters are the storage form of the vitamin in the liver. Specific hydrolytic and esterification enzymes occur in liver, intestine and in the retina of the eye.

The major forms of vitamin A ingested by man are the esters of retinol with long chain fatty acids. Compared with the normal pattern of fatty acids in tissue lipids, they are a relatively select group and much more saturated. Palmitic acid predominates, together with smaller quantities of stearic and oleic acids. When the esters reach the lumen of the small intestine, they are almost completely hydrolysed, absorbed into the intestinal cells and re-esterified. They are transported as components of chylomicrons to the liver where they are stored almost entirely as retinyl palmitate regardless of the composition of the dietary esters. Modification of dietary retinyl esters to such a specific fatty acid composition requires, as in the case of the cholesterol esters, specific hydro-

lytic and synthetic enzymes; these occur mainly in the intestine, liver and in the retina of the eye.

There are two forms of the hydrolytic enzyme, one of which is specific for short chain esters like retinyl acetate even though this ester does not occur naturally! The other has maximum activity with retinyl palmitate as substrate but also hydrolyses other long chain esters. As in the hydrolysis of cholesterol esters, the enzyme is not just a non-specific *esterase*, but has quite definite specificity for retinyl esters. In vitamin A deficiency, the activity of the enzyme increases one hundred fold.

The esterification enzyme resembles the 'low energy' cholesterol esterase in that neither ATP nor CoA appear to take part in the reaction. Nor are free fatty acids or acyl-CoA thiol esters incorporated into retinyl esters. One of the major problems in this area of research is to identify the acyl donor. One suggestion is that, as in plasma cholesterol ester biosynthesis, the donor is a phospholipid. This possibility still remains to be tested.

WAXES

Strictly speaking, the word *wax* should be reserved for the esters of long chain fatty acids with long chain primary alcohols. These substances generally occur in nature at the surfaces of animals and plants as part of a mixture with long chain hydrocarbons and the free alcohols and fatty acids. Consequently the name wax is often used for the whole surface lipid mixture rather than specifically for the

esters. In this section we shall discuss not only the esters, but also say something about the hydrocarbons because their biosynthesis involves stepwise condensations of fatty acids and only at the final stage is the carboxyl group removed to form a hydrocarbon.

The plant and animal waxes are responsible for the water repellent character of the surface and are therefore important in conserving the organism's own water and providing a *barrier* against its environment. 'True' waxes are also found in *mycobacteria* and *corynebacteria*, as well as *waxes A,B,C and D* which are really complex esters of the hydroxy fatty acid mycolic acid with a peptidopolysaccharide (see chapter 6).

'True' waxes are esters of long chain (10-30 carbon atoms) alcohols with long chain fatty acids.

The component fatty acids and alcohols of the majority of waxy esters have an even number of between 10 and 30 carbon atoms in a straight chain. Some are more complex and may contain all the various types of fatty acids described in chapter 2, such as those with branched chains, unsaturation or ring systems.

The biosynthesis of these waxy esters is a relatively new area of study. P.E. Kolattukudy in the U.S.A. has suggested that all three mechanisms which are known to exist for the biosynthesis of cholesterol esters, namely (*i*) reversal of esterase activity, (*ii*) transfer of acyl groups from acyl-CoA and (*iii*) transfer of acyl groups from a phospholipid, also hold good for wax biosynthesis. However, although there is suggestive evidence for all three mechanisms, none seems to have been established beyond doubt.

The long chain hydrocarbon components of wax are formed by elongation of long chain fatty acids followed by decarboxylation.

The hydrocarbon components of the surface waxes are usually *n*-paraffins with an odd number of between 21 and 35 carbon atoms. While the proportion of alkanes in the mammal waxes is usually less than one per cent, in plant surface lipids it may be as much as 50% and in insects even as high as 75%. One hydrocarbon is usually predominant. Many theories have been put forward about their biosynthesis, tested and found wanting. At the present time, most of the evidence points to the elongation of an existing long chain fatty acid (usually $C_{16:0}$ or $C_{18:0}$) by successive additions of two-carbon (or perhaps larger) units. When the chain length reaches about 30 carbon atoms, the fatty acid is decarboxylated to yield a C_{29} paraffin, hence the occurrence of only odd chain-length hydrocarbons.

The so-called *elongation-decarboxylation* pathway for hydrocarbons is quite distinct from the normal fatty acid synthetase. Palmitic and stearic acids are much more rapidly converted into long chain hydrocarbons than acids with chain lengths shorter than 16. In plants, the absence of light, which inhibits the incorporation of acetate into fatty acids, has little effect on the biosynthesis of paraffins provided that precursor fatty acids are available. A favourite method by which biochemists distinguish between different pathways is by using specific inhibitors. Chlorophenyldimethylurea, for example, inhibits the incorporation of acetate into fatty acids but has no effect on hydrocarbon biosynthesis. On the other hand, trichloroacetate at low concentrations strongly inhibits hydrocarbon biosynthesis, but does not affect fatty acid synthesis.

The intermediates on the pathway to long chain hydrocarbon synthesis are very long chain fatty acids, and at various stages of the pathway ($C_{20} - C_{28}$) some of these may get chanelled off and esterified into glycerides or phospholipids. The production of these very long chain fatty acids is affected by the same factors as paraffin biosynthesis, arguing for a distinct mechanism for normal and very long chain fatty acid biosynthesis.

Analytical techniques

Owing to the similar physical properties of different types of neutral lipids, a combination of separation techniques is necessary to resolve complex natural mixtures.

The properties of neutral lipids are largely those of their acyl chains and therefore many of the methods of fatty acid separation which we described in chapters 1 and 2 are equally applicable to a whole variety of neutral lipids. By definition, neutral lipids tend to have similar physical properties and as chromatographic techniques are largely dependent on these physical characteristics, then a careful choice and combination of methods is necessary to achieve full resolution. Chromatography on columns (for large quantities) and thin layers (for small or medium quantities) of silicic acid or alumina are most often used. In addition, 'Florisil' (Magnesium silicate) is especially suitable for neutral lipids because of its fast flow rate and because polar lipids stick tightly and are not eluted. By impregnating the stationary phase with different substances,

additional resolutions can be achieved which would be impossible on plain silica or alumina. Compounds with different degrees of unsaturation can be separated if the impregnating substance is silver nitrate, while compounds of different chain length are resolved by 'reversed phase' chromatography. Gas—liquid chromatography is being used more and more for substances other than fatty acid methyl esters. Substances like glycerides and sterol esters are much less volatile than fatty acid methyl esters, so that the temperature at which the separation is achieved has to be much higher, adding to the risk of degradation. Another approach in the case of non-volatile hydroxy compounds is to make derivatives which have lower boiling points such as trimethylsilyl ethers and to analyse these by GLC.

Sterol esters and retinyl esters are much less polar than their parent alcohols so that ester and alcohol can be quite easily separated by TLC or column chromatography on silicic acid or alumina provided that a solvent of appropriate polarity is chosen. Once the mixtures have been resolved and the components purified, the methods of quantitative determination are widely different for each type of neutral lipid. Glycerides are usually determined by measuring the number of fatty ester bonds by the 'hydroxamate method'. The ester is treated with alkaline hydroxylamine, which reacts specifically with ester-linked fatty acids but not with non-esterified fatty acids. The resulting hydroxamate salts form a very characteristic brown complex with ferric chloride, the concentration of which can be measured by spectrophotometry.

Sterols are specifically precipitated by digitonin and the resulting sterol digitonide can be determined spectrophotometrically by the colour which it develops with a solution of anthrone in acetic acid. The conjugated double bond system characteristic of vitamin A provides a sensitive means of quantitative analysis — namely measurement of the fluorescence. The molecule is excited at a wavelength of 325 nm and the light emission measured at 470 nm.

SUMMARY

The term *neutral* lipid is used to describe a rather large number of compounds differing widely in their chemical properties but having similar physical characteristics. They have no ionized or polar groups and therefore tend to be extracted more easily into apolar solvents and to 'run faster' on chromatograms in a given solvent than do *polar* lipids.

Glycerides are esters of one, two or three fatty acids with glycerol. Different combinations of the very large number of naturally occurring fatty acids can give rise to a very large number of individual glyceride molecules although particular species of animals or plants have characteristic glyceride patterns. Recent work has proved that the glyceride fatty acids in many species are not randomly distributed but that each position on the glycerol backbone tends to be esterified with specific types of fatty acids. Methods of analysing the fatty acids on each position have been developed. Glycerides are built up either *de novo* by a pathway involving acylation of glycerol phosphate, or by acylation of monoglycerides, a pathway by which dietary glycerides are modified. Triglycerides serve as a very convenient anhydrous storage form of fatty acids in adipose tissue. The release of glyceride fatty acids for purposes of oxidation and energy production is brought about by lipases under fine hormonal control. Other lipases, released from the pancreas into the intestinal lumen and from the capillaries into the blood plasma also control the hydrolysis of dietary and plasma triglycerides respectively.

Sterol esters are the predominant forms of steroids in many tissues such as plasma, liver and the adrenal gland. Hydrolysis and re-esterification are continually occurring and both reactions may be catalysed by the enzyme sterol ester hydrolase. Other enzyme activities synthesize cholesterol esters by transfer of fatty acids from acyl-CoA or phosphatidyl choline to free cholesterol.

Less is known about the enzymes cleaving retinyl esters or synthesizing them from retinol and fatty acids. Enzymes analogous to those involved in sterol ester synthesis may be involved.

Waxes are found in the surface lipids of plants or animals. True waxes are esters of long chain fatty acids with long chain alcohols. Often included in the term *wax* are the long chain (odd numbered) alkanes which are formed by condensation of normal long chain fatty acids with two-carbon units followed by decarboxylation at about the C_{30} level.

A combination of chromatographic techniques is required to separate complex natural mixtures of neutral lipids which have such similar physical properties. These include GLC, TLC, and column chromatography. Additional resolution is achieved by impregnation of the stationary phase with different substances, as in *argentation* chromatography or *reversed phase* chromatography.

BIBLIOGRAPHY

Glycerides

1. SENIOR, J.R. (1964). 'Intestinal absorption of fats'. *J. Lipid Res.* 5, 495.

2. DAWSON, A.M. (1967). 'Absorption of Fats' *Brit. Med. Bull.* 23, 247.

3. JOHNSTON, J.M. (1968). 'The Mechanism of Fat Absorption' in C.F. Code and W. Heidel (ed.) Handbook of Physiology, Section 6 The Alimentary Canal, Vol III, Intestinal Absorption, Williams and Wilkins Co. Baltimore.

4. HÜBSCHER, G.H. (1970). 'Glyceride Metabolism' in S.J. Wakil (ed.) *Lipid Metabolism*, Academic Press Inc, New York, p. 280.

5. COLEMAN M.H. (1963). 'The structural investigation of natural fats'. *Advances in Lipid Research* 1, 2.

6. MASORO E.J. (1968). 'Physiological chemistry of lipids in mammals'. B. Saunders, Philadelphia.

7. ROBINSON D.S. (1963). 'The clearing factor lipase and its action in the transport of fatty acids between blood and tissues.' *Advances in Lipid Research* 1, 34.

Cholesterol Esters

8. GOODMAN D.S. (1965). 'Cholesterol ester metabolism, *Physiol. Rev.* 45, 747.

9. GLOMSET J.A. (1968). The plasma lecithin cholesterol acyl transferase reaction. *J. Lipid Res.* 9, 155.

Waxes

10. KOLATTUKUDY P.E. (1968). Biosynthesis of surface lipids. *Science, N.Y.* 159 498.

Vitamin A esters

11. OLSON J.A. (1967). The metabolism of vitamin A, *Pharmacol. Rev.* 19, 559.

4 Phospholipids

We began this book by considering the gross heterogeneity of chemical structures covered by the term *lipid*. Although at first sight, the name *phospholipid* might seem to depict a narrower and more closely related group of compounds, on closer investigation the variety of molecules seems little less bewildering. Indeed the presence of a phosphate group and a fatty acid chain seems to be the nearest one can get to a definition. Even the name changes according to the author, and *phosphatide* is a commonly found alternative. Perhaps we may take J.N. Hawthorne's classification as a convenient starting point for attempting a survey of the different phospholipid types, and make two large groupings according to the alcohol with which the fatty acids and phosphoric acid are esterified: (1) Glycerophospholipids (glycerophosphatides, phosphoglycerides)

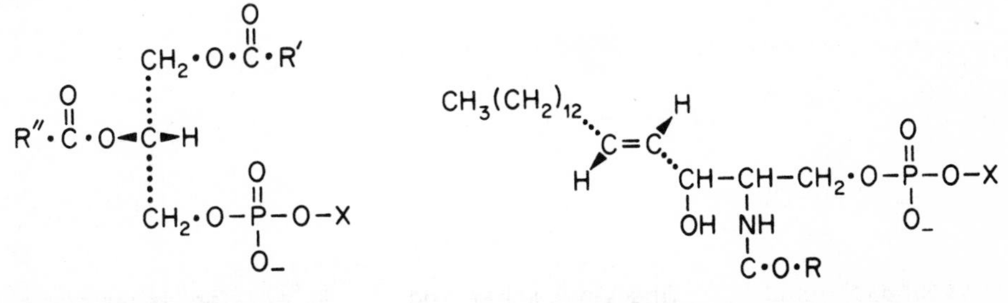

(1) Glycerophospholipids

(2) Sphingophospholipids

Fig. 4.1. Classes of phospholipids.

having glycerol as the alcohol;
(2) Sphingophospholipids (Phosphosphingolipids) having sphingosine as the alcohol (Fig. 4.1).

The almost infinite variety of glycerophospholipids.

The group derived from glycerol is by far the most abundant and phospholipids of this type occur in every living cell. The group can be further subdivided into classes according to (a) the group 'X' and (b) the type of carbon chain and the kinds of bond by which that chain is linked to the glycerol. An idea of the wide spectrum of molecular types can be appreciated from Table 4.1. These compounds can be thought of as derivatives of glycerides in which the hydroxyl on carbon atom 3 is esterified with phosphoric acid which in its turn is esterified with a range of small molecules – organic bases, amino acids, alcohols, represented by 'X'. Alternatively, as we pointed out in chapter 1, the naming system

phosphatidyl-X implies that they are derivatives of glycerol-3-phosphate, an approach which may be more helpful later when considering their biosynthesis. Inevitably, there is 'overlap' between the lipids of different classes in any arbitrary classification scheme and some of the phospholipids in Table 4.1 may equally well be considered as *glycolipids*. However, any lipid possessing a phosphate group will here be considered as a *phospholipid* whether or not it contains a sugar.

Even more possibilities exist for variations in the acyl groups. Animal phospholipids contain mostly fatty acids between chain length 16 and 20 with palmitic (16:0), stearic (18:0), oleic (18:1), linoleic (18:2) and arachidonic (20:4) predominant. Plant leaf phospholipids have a more limited range with very few fatty acids greater than C_{18}. In the leaf, palmitic, oleic and linoleic predominate while the very abundant α–linolenic acid ($^{\Delta 9,12,15}C_{18:3}$) accumulates mainly in the galactolipids rather than in the phospholipids. It is when we turn to the micro-organisms that the fatty acid picture becomes so complex. Bacteria

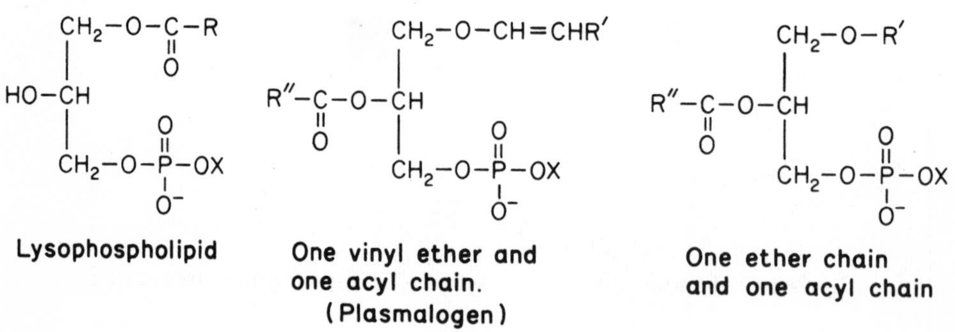

Fig. 4.2. Variations in the hydrocarbon chain.

Table 4.1. Structural variety of diacyl-glycerophospholipids

X	Name of Phospholipid	Source	Remarks
H	Phosphatidic acid	Animals, higher plants, micro-organisms.	Only minute amounts found. Main importance as a biosynthetic intermediate
$OH \cdot CH_2CH_2 \overset{+}{N}(CH_3)_3$ Choline	Phosphatidyl choline (lecithin)	Animals. First isolated from egg yolks. Higher plants. Rare in micro-organisms.	Most abundant animal phospholipid
$OH \cdot CH_2CH_2NH_2$ Ethanolamine	Phosphatidyl ethanolamine	Animals, higher plants, micro-organisms.	Widely distributed and abundant. Major component of old 'cephalin' fraction. N-acetyl derivatives in brain; fatty amides in wheat flour, peas.
$OH \cdot CH_2CH \cdot NH_2$ Serine |—COOH	Phosphatidyl serine	Animals, higher plants, micro-organisms	Widely distributed but in small amounts. Minor component of old 'cephalin' fraction. Serine as L isomer. Lipid usually in salt form with K^+, Na^+, Ca^{++}.
Myo-inositol	Phosphatidyl inositol	Animals, higher plants, micro-organisms.	The natural lipid is found as a derivative of myo-inositol-1-phosphate only.
Inositol-4-phosphate	Phosphatidyl inositol phosphate. (Diphosphoinositide)	Animals, trace in yeast.	Mainly nervous tissue but minute amounts in other tissues, esp. plasma membrane
Inositol-4,5-diphosphate	Phosphatidyl inositol diphosphate. (Triphosphoinositide).	Animals, trace in yeast	Mainly nervous tissue but minute amounts in other tissues, esp. plasma membranes.

Table 4.1 (continued)

X	Name of Phospholipid	Source	Remarks
 O-mannose Inositol mannoside (ring structure with positions 1–6, OH groups, HO substituents) O-(mannose)	Phosphatidyl inositol mannoside $x = 0$, monomannoside $x = 1$, dimannoside etc.	Micro-organisms (*M.phlei*, *M. tuberculosis*)	
$CH_2OH \cdot CHOH \cdot CH_2OH$ Glycerol	Phosphatidyl glycerol	Mainly higher plants and micro-organisms	'Free' glycerol has opposite stereochemical configuration to the acylated glycerol, i.e. 1,2-diacyl-*sn*-glycero-3-phosphoryl-1-*sn*-glycerol. Probably most abundant phospholipid.
$\quad\;\; O\;\; NH_2$ $\quad\;\; \| \quad \|$ $CH_2O \cdot C \cdot CH \cdot R$ Aminoacyl glycerol $-CHOH$ $-CH_2OH$	Aminoacyl phosphatidyl glycerol	Micro-organisms	Amino acids mainly *L*-lysine and *L*-alanine but also ornithine and glycine found. Probably esterified to C(3) of 'free' glycerol.
Glucosaminyl glycerol	Glucosaminyl phosphatidyl glycerol	Micro-organisms	*D*-glucosamine linked by glycosidic bond to C(2) or C(3) of 'free' glycerol
Glucose	Phosphatidyl glucose		Glucofuranose form

have no polyunsaturated fatty acids but make up for this with a wide variety of short chain saturated fatty acids, branched chain fatty acids and cyclopropane derivatives many of which are esterified in phospholipids.

As we shall see later, most cells contain enzymes which can remove either or both of the phospholipid fatty acids. The monoacyl compound remaining after the action of the 'phospholipase' is called a *lysophospholipid* (Fig. 4.2). Most cells contain small concentrations of these compounds but the levels cannot build up owing to their strong detergent properties. The name *lyso-* derives from their ability to 'lyse' or disrupt red blood cell membranes by detergent action and this property has been used to estimate concentration of lysophospholipids.

Not all phospholipids have their hydrocarbon residues linked exclusively by an ester bond to glycerol. *Plasmalogens* posses a vinyl ether linkage and this probably always at the 1-position. The bases esterified to the phosphate are usually choline or ethanolamine but a serine derivative has been described. Most animal tissues contain rather small

amounts but they are abundant in nervous tissue and heart muscle mitochondria. Compounds containing a normal ether linkage are much rarer, occurring in erythrocytes of some species and in slugs (Fig. 4.2). An unusual diether phospholipid occurs in *Halobacterium cutirubrum* (Fig. 4.3). This brief account of phospholipid structures does not claim to be exhaustive and references to further reading will be found at the end of the chapter, but perhaps we should mention a class of phosphorus-containing lipids which is having a lot of attention these days, namely the *phosphonolipids*. These have a carbon-phosphorus bond instead of the more usual P-O-C bond (Fig. 4.4).

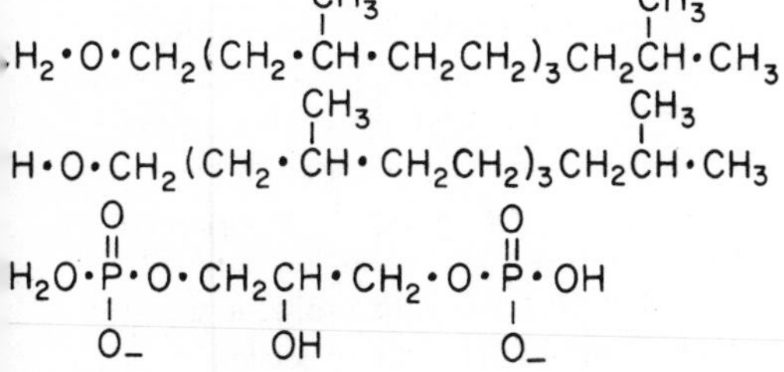

$$
\begin{array}{l}
\text{CH}_2\text{-O-C-R}' \\
\quad\quad\quad \| \\
\text{R}''\text{-C-O-CH} \quad\quad \text{O} \\
\quad \| \quad\quad\quad | \\
\quad \text{O} \quad\quad\quad \text{CH}_2\text{-O-P-CH}_2\text{-CH}_2\text{-NH}_3^+ \\
\quad\quad\quad\quad\quad\quad | \\
\quad\quad\quad\quad\quad\quad \text{O}^-
\end{array}
$$

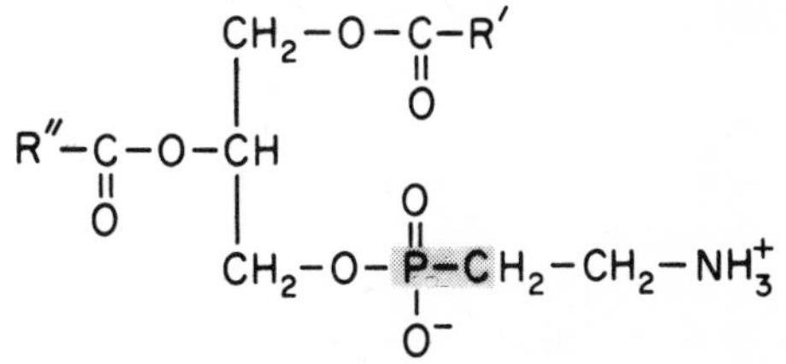

Fig. 4.4. Phosphonolipid.

Some phospholipids are derived from sphingosine instead of glycerol.

A rather less abundant but nevertheless important group of phospholipids has sphingosine as its alcoholic moiety. (Figs. 4.1 and 5.2). Group 'X' is usually choline (*Sphingomyelin*) and there is less possibility for variation as there is only one fatty acid chain. The most significant difference from glycerolipids is that the fatty acid residue is here linked to an amino group. The alcohol itself is subject

$$
\begin{array}{l}
\quad\quad\quad\quad\text{CH}_3 \qu\quad\quad\quad\quad\quad \text{CH}_3 \\
\quad\quad\quad\quad | \qu\quad\quad\quad\quad\quad\quad | \\
\text{H}_2\cdot\text{O}\cdot\text{CH}_2(\text{CH}_2\cdot\text{CH}\cdot\text{CH}_2\text{CH}_2)_3\text{CH}_2\text{CH}\cdot\text{CH}_3 \\
\qu\quad\quad\quad\text{CH}_3 \ququad\quad\quad\quad \text{CH}_3 \\
\ququad\quad\quad | \quad\quad\quad\quad\quad\quad | \\
\text{H}\cdot\text{O}\cdot\text{CH}_2(\text{CH}_2\cdot\text{CH}\cdot\text{CH}_2\text{CH}_2)_3\text{CH}_2\text{CH}\cdot\text{CH}_3 \\
\quad\text{O} \quad\quad\quad\quad\quad\quad\quad \text{O} \\
\quad \| \quad\quad\quad\quad\quad\quad\quad\quad \| \\
\text{H}_2\text{O}\cdot\text{P}\cdot\text{O}\cdot\text{CH}_2\text{CH}\cdot\text{CH}_2\cdot\text{O}\cdot\text{P}\cdot\text{OH} \\
\quad | \quad\quad\quad\quad | \quad\quad\quad\quad | \\
\quad \text{O}_- \quad\quad\quad \text{OH} \quad\quad\quad \text{O}_-
\end{array}
$$

Fig. 4.3. Diether phospholipids of Halobacterium.

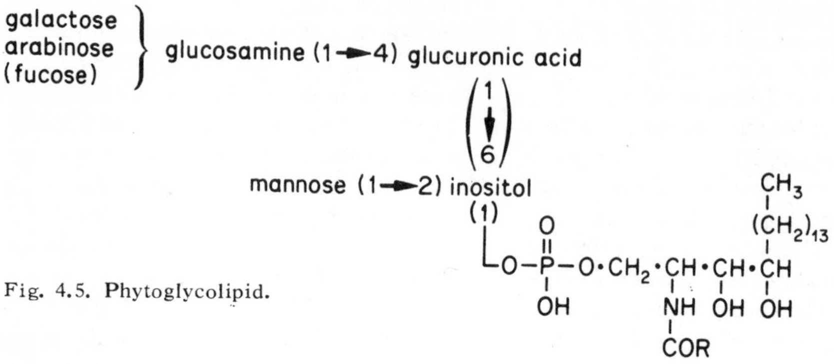

galactose ⎫
arabinose ⎬ glucosamine (1→4) glucuronic acid
(fucose) ⎭

$$
\begin{pmatrix} 1 \\ \downarrow \\ 6 \end{pmatrix}
$$

mannose (1→2) inositol
(1)

$$
\underset{OH}{\overset{O}{\underset{\|}{L_O-P}}}-O \cdot CH_2 \cdot \underset{NH}{CH} \cdot \underset{OH}{CH} \cdot \underset{OH}{CH} \cdot (CH_2)_{13} \cdot CH_3
$$

NH
|
COR

Fig. 4.5. Phytoglycolipid.

to modification (see also Chapter 5); probably the most interesting variation being found as phytosphingosine in the lipid phytoglycolipid. This is exclusively a plant lipid, originally isolated from corn and soya bean phospholipids but also found in wheat germ, flax, cotton and sunflower seeds. The terminal sequence of monosaccharides is still to be determined as indicated in Fig. 4.5.

It will be apparent from this brief summary that the range of structural types of phospholipids is almost limitless. Even if the natural products chemist isolates a phospholipid with a given base (which has formerly been the governing factor in any separation method), that product will still contain a large number of 'species' of different fatty acid composition. Milk fat, for example, contains about 95 different phosphatidyl cholines! Recently, as will be described later, it has become possible to separate these species and therefore it is no longer

sufficient to talk about, say, 'phosphatidylcholine' but individual species of 'phosphatidyl cholines', for example 1-palmitoyl-2-stearoyl-phosphatidylcholine.

Phospholipids were once thought to be metabolically inert.

So far. we have described what are essentially the results of the work of the natural products chemist who has isolated phospholipids from living tissues and worked out their chemical structures. The tools which he uses to isolate and identify these structures will be discussed at the end of this chapter. Our main concern here is the way in which they originate in the organism. An interest in the *chemistry* of phospholipids began with the quite extensive investigations of J.L.W. Thudichum, who isolated and analysed lipids from many animal tissues, particularly brain and published his results in a now historical document, 'A Treatise on the Chemical Constitution of the Brain' (1884). Later it began to be appreciated that the difficulties involved in handling these substances and obtaining a 'pure' product were enormous. Another factor tending to discourage research into phospholipids was the very prevalent but erroneous idea that they were metabolically inert and that once laid down during the initial growth of the tissue their 'turnover' was very slight and hence a purely structural significance was ascribed to them.

'Tracer' studies revolutionized concepts about biosynthesis.

This myth was exploded by the Danish chemist George Hevesy who in 1935 demonstrated that the radioactive isotope of phosphorus (^{32}P) (recently discovered by Joliot and Curie) could be rapidly incorporated as inorganic orthophosphate into tissue phospholipids. Previously, the chemist Schoenheimer, working at Columbia University, New York, had also demonstrated the incorporation of the stable isotopes (^{2}H) and (^{15}N) into proteins and fats — findings which revolutionized our concepts of metabolism. From these studies emerged the concept of 'turnover' meaning the continual replacement of molecules or *parts* of molecules by synthesis and breakdown. Another important concept to emerge from research involving the tracer method was that of the 'metabolic pool.' This is a circulating mixture of chemical substances in partial or total equilibrium with similar substances derived by release from cellular tissues or from the diet, which the organism can use for the synthesis of new tissue constituents.

That tissues are able to synthesize their own phospholipids could also be inferred from the fact that they are not essential dietary requirements, but in spite of the impetus given to biosynthetic studies by the tracer work of the 1930's and 1940's, the major treatise on phospholipids in the early '50's (Wittcoff's 'Phosphatides') could give virtually no information about their biosynthesis.

A mechanism is found for the biosynthesis of the parent phospholipid, phosphatidic acid.

As the various components of the phospholipid molecule — fatty acids, alcohol, phosphate and base 'turn over' independently, we must first know the origin of each constituent and then learn how they are welded together. Two years after the publication of 'The Phosphatides' and 18 years after Hevesy's demonstration of the rapid rate of turnover of phospholipids came the first real breakthrough in understanding how the complete phospholipid molecules are built up. Two American biochemists, A. Kornberg and W.E. Pricer, found that a cell-free enzyme preparation from liver would 'activate' long chain fatty acids by forming their coenzyme A esters (we have already shown in Chapter 2 how the fatty acids themselves are produced by the *synthetase*). Then they went on to demonstrate that these activated fatty acids, catalysed by an *acyl transferase*, would esterify 3-*sn*-glycerophosphate to form 1,2-diacyl-3-*sn*-glycerophosphate (phosphatidic acid). This is the parent molecule of all the glycerophospholipid family and was at first thought not to be a 'normal constituent' of tissue lipids. Later studies have shown it to be widely distributed but in minute amounts. Thus glycerophosphate is one of the building units for phospholipid biosynthesis. From which metabolic pathway is it derived? The major proportion is believed to originate from one of the chief pathways of carbohydrate breakdown — the so-called glycolysis or Emden-Meyerhof pathway. Reduction of dihydroxy acetone phosphate, a key intermediate in this sequence, yields 3-sn-glycerol phosphate and this provides another link between carbohydrate and lipid metabolism. (Recently a new lipid, *acyl dihydroxyacetone phosphate*, has been discovered, theoretically providing for a new entry into phospholipid metabolism as an alternative to phosphatidic acid, but its quantitative significance has not yet been elucidated). Alternative sources of glycerophosphate are also available, as for example the asymmetric phosphorylation of the apparently symmetrical molecule glycerol, catalysed by the enzyme *glycerol kinase* to yield exclusively 3-sn-glycerophosphate. (The natural compound always has the '3-sn-' structure and therefore in future we can omit the prefix and simply use the word *glycerophosphate*).

Kornberg and Pricer's mechanism, however, is not the exclusive route to the simplest phospholipid. Another pathway involves the phosphorylation of a diglyceride by ATP in the presence of the enzyme *diglyceride kinase*. Quite recently, stimulated by the upsurge of interest in the acyl carrier protein, the American, H. Goldfine, working at Harvard University proved that certain bacteria employed acyl-ACP as the donor of fatty acids to glycerophosphate, rather than the CoA derivatives. The mechanism in higher plants has not been thoroughly investigated.

A novel cofactor for phospholipid synthesis was found 'by accident'.

After the problem of phosphatidic acid biosynthesis had been solved, interest then grew in the pathways to more complex lipids. What was the origin of the group 'X'? The first significant finding in connection with phosphatidyl choline biosynthesis was made when Kornberg and Pricer demonstrated that the molecule *phosphoryl choline* was incorporated *intact* into the lipid. This they did by incubating phosphoryl choline, 'labelled' both with ^{32}P and ^{14}C in a known proportion, with a liver enzyme preparation and finding that the ratio of ^{32}P to ^{14}C radioactivities remained the same in phosphatidyl choline. Choline may come *via* a number of pathways having their origins in protein metabolism. For instance the amino acid serine may be decarboxylated to ethanolamine, which may then be methylated to form choline. Phosphoryl choline arises by phosphorylation of choline with ATP in the presence of the enzyme *choline kinase*. The story of how the base cytidine, familiar to nucleic acid chemists, was found to be involved in complex lipid formation is a good example of how some major advances in science are stumbled upon by accident, although the subsequent exploitation of this finding by the American biochemist, E.P. Kennedy, typifies careful scientific investigation at its best. Kennedy proved that cytidine triphosphate (CTP) which was present as a small contaminant in a

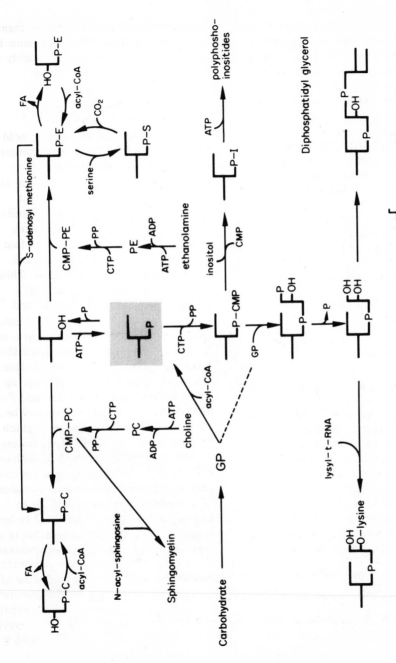

Fig. 4.6. The biosynthesis of some important phospholipids. $\vdash$P–X = phosphatidyl–X; C = choline; E = ethanolamine; I = inositol; S = serine; CMP–P–C = cytidine diphosphocholine, etc; GP = glycerol–3–phosphate; FA = fatty acid; P = O=P–(OH)₃; PC,PE = phosphoryl choline, ethanolamine.

sample of ATP was the essential cofactor involved in the incorporation of phosphoryl choline into the lipid and later isolated the 'active form of phosphoryl choline', namely cytidine diphosphocholine (CDP-choline). The adenine analogue had no reactivity.

Two enzymes are involved in the sequence (See Fig. 4.6). One transfers phosphoryl choline to cytidine monophosphate with release of pyrophosphate (EC.2.7.7.15, *CTP : choline phosphate cytidyl transferase*); the other transfers phosphoryl choline from the nucleotide coenzyme to diglyceride to form the phospholipid, (EC.2.7.8.2; *1,2-diglyceride-choline phosphotransferase*). CTP can be regenerated by phosphorylation of the CMP released in the second reaction; the phosphate donor is ATP.

Diglyceride arises from degradation of phosphatidic acid.

Further investigation into the origin of the diglyceride needed for this reaction revealed that phosphatidic acid is indeed involved in the synthesis of more complex lipids. The enzyme *phosphatidate phosphohydrolase* is widely distributed in tissues and its function is to split off the inorganic phosphate from phosphatidic acid to yield a diglyceride. It is this diglyceride which acts as an acceptor for the 'activated phosphoryl choline'. This fact explains the extremely high 'turnover' which is almost always observed in the phosphatidic acid fraction of tissues which have been exposed to ^{32}P. This lipid is being continually synthesized and broken down to provide a basis for the accumulation of the more abundant phospholipids. In plasmalogen biosynthesis the *acceptor* to which phosphoryl choline is transferred from CDP-choline is *plasmalogenic diglyceride* or *1-alkenyl-2-acyl-sn-glycerol* (chapter 3, fig. 3.2).

Two alternative pathways are available for the biosynthesis of sphingomyelin.

The 'cytidine nucleotide pathway' operates in an almost exactly analogous way for the formation of phosphatidyl ethanolamine and sphingomyelin. In the formation of sphingomyelin, the *acceptor* for phosphoryl choline is of course not a diglyceride but *N*-acyl sphingosine or *ceramide* and the enzyme is *phosphoryl choline-ceramide transferase*, (Reaction 1, below). An alternative biosynthetic route to sphingomyelin, which does not have a close analogy in the case of phosphatidyl choline is the acylation of sphingosylphosphoryl choline by acyl-S.CoA, (Reaction 2b, below). Sphingosylphosphoryl choline is formed by transfer of phosphoryl choline from CDP-choline to free sphingosine (Reaction 2a). The relative importance of these two pathways *in vivo* has not been established.

(1)   N-acyl-sphingosine + CDP-choline ⟶ Sphingomyelin + CMP

(2a) Sphingosine + CDP-choline ⟶ Sphingosylphosphoryl choline + CMP

(2b) Sphingosylphosphoryl choline + acyl-S-CoA ⟶ Sphingomyelin + CoASH.

*In the biosynthesis of another
group of phospholipids the cytidine
nucleotide is CDP-diglyceride, not
CDP-base.*

The 'cytidine nucleotide pathway' undergoes an important modification in the case
of certain other lipids such as phosphatidyl
inositol and phosphatidyl glycerol. For
these lipids, the cytidine coenzyme is in
a *lipid-soluble* rather than a water-soluble
form, namely as CDP-diglyceride which is

formed from phosphatidic acid by transfer
of CMP from CTP. This 'active
phosphatidic acid' can then be transferred
to free inositol by another enzyme
illustrated in Fig. 4.7.
The more complex *phosphoinositides* which
contain further phosphate groups at
positions 4 and 5 in the inositol ring
(See Table 4.1) can be formed by brain
and liver enzymes with ATP as the
phosphate donor. CDP-diglyceride can

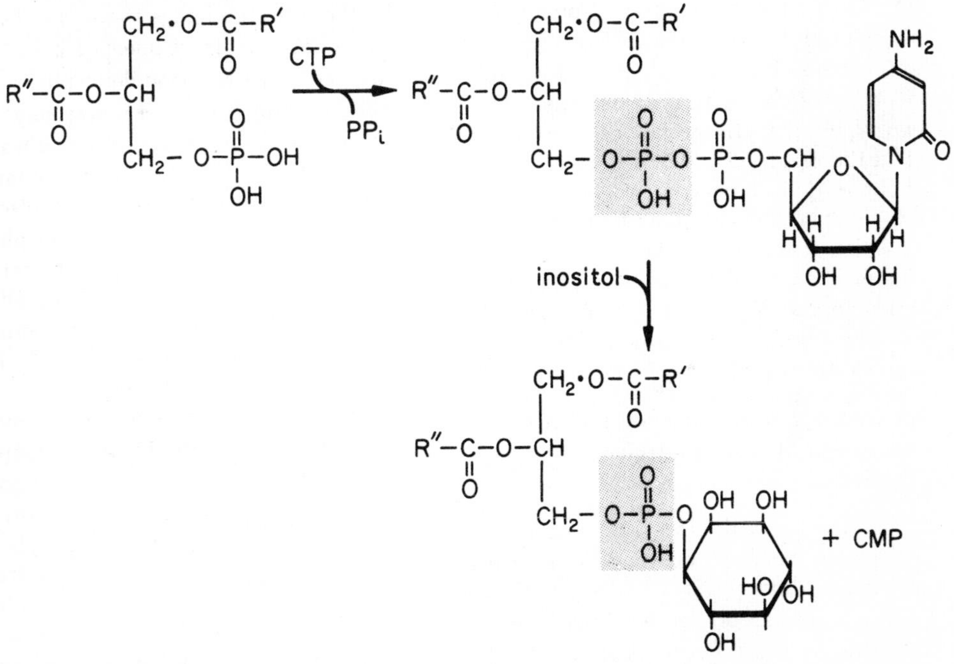

Fig. 4.7. Biosynthesis of phosphatidyl inositol.

also act as an intermediate in the formation of phosphatidyl glycerol by the transfer of a phosphatidyl group to glycerophosphate, which is then 'dephosphorylated' (Fig. 4.6). This reaction has been demonstrated by Kennedy with cell-free extracts from the bacterium *Escherichia coli*. A consequence of this mechanism is that the 'free' (or non-acylated) glycerol has the opposite stereochemical configuration (i.e. 1-*sn*-glycerol phosphate) to the first glycerol. Diphosphatidyl glycerol (Cardiolipin) is formed by a second transfer of a phosphatidyl group from CDP-diglyceride to phosphatidyl glycerol. Probably the most significant difference between the two types of 'cytidine nucleotide pathway' is in the origin of the phosphorus. In phospholipids formed from 'CDP-base', the phosphorus originates from the phosphorylated base, and not from phosphatidic acid, whereas in phospholipids formed by way of CDP-diglyceride, the phosphorus arises *directly* from phosphatidic acid.

Some phospholipids may be formed from others by a 'base exchange' reaction

In comparison with other lipids, very little work has been done on the synthesis of phosphatidyl serine. The biosynthesis of this lipid has presented something of a puzzle as it did not at first appear to be formed by either of the cytidine nucleotide pathways. Recent studies have indicated that both phosphatidyl serine and phosphatidyl ethanolamine are formed by way of the 'CDP-diglyceride' pathway in the bacterium *E.coli*. The only pathway which has been definitely proved to occur in animal tissues however is an

'exchange reaction' catalysed by an enzyme from a liver microsomal fraction in the presence of Ca^{++} ions. In the process, the ethanolamine of phosphatidyl ethanolamine may be replaced by serine. Unlike the other pathways, the reaction requires no input of energy from 'energy-rich molecules' such as ATP. It resembles the reverse of the reaction in which the base is split off by phospholipase D (see the later section on *Phospholipases*) but although some phospholipids are undoubtedly formed by the base exchange catalysed by phospholipase D, it seems likely that the exchange reaction described here is catalysed by a separate enzyme. There is an alternative route to phosphatidyl ethanolamine from phosphatidyl serine because serine can be decarboxylated in the lipid form. This pathway is not considered to be of great importance because in most studies *in vivo* (that is 'whole, living animal' studies as opposed to *in vitro* or 'test tube' preparations) the labelling of phosphatidyl serine from labelled inorganic phosphate is very slow compared with that of phosphatidyl ethanolamine.

Another way of forming phosphatidyl choline is by enzymic methylation of ethanolamine in phosphatidyl ethanolamine

Phosphatidyl ethanolamine itself can form the basis for yet another alternative pathway to phosphatidyl choline. The first clue to this pathway was the discovery that phospholipids from certain sources contain small amounts of the bases mono- and dimethylaminoethanol. Later, enzymes from the mould *Neurospora* and from liver were discovered which transfer methyl groups from the amino acid methionine to the amino group of phosphatidyl ethanolamine. The 'active form' of methionine is the molecule S-adenosyl methionine, which, as we have seen in Chapter 2, is the donor of the ring methylene group in cyclopropane fatty acids. Only a few rather primitive bacteria contain phosphatidyl choline, and in these the major biosynthetic route appears to be by methylation of phosphatidyl ethanolamine rather than *via* CDP-choline, in contrast to most animal tissues.

The fatty acid patterns of phospholipids may be 'tailored' by independent exchange with fatty acids on preformed lipids.

Earlier, we discussed the concept of the independent 'turnover' of individual parts of the phospholipid molecule. We have seen that phosphate may be incorporated via glycerophosphate or phosphorylation of diglyceride, and eliminated by phosphatidate phosphohydrolase. The base may often be incorporated and removed by an exchange reaction. Do fatty acids have an independent turnover and how do certain phospholipids attain their quite distinctive fatty acid patterns? Does the fatty acid composition arise from the specificity of the enzymes which acylate the parent molecule to give phosphatidic acid (according to the Kornberg and Pricer mechanism), or at a later stage? The answers to these questions began to come from the series of very thorough researches (beginning in the early 1960's and still continuing) by the American biochemist W.E.M. Lands, working at Ann Arbor, Michigan. He incubated 2-lysophosphatidyl choline and a series of different acyl-CoA esters with a microsomal fraction from liver and found that the corresponding diacyl phospholipids were formed, the rates of acylation depending on the types of fatty acids incorporated. By and large, unsaturated fatty acids were more readily incorporated into a 2-lyso-compound by this enzyme than saturated acids. Lands knew that the tissue contained an enzyme (phospholipase A) capable of hydrolysing the fatty acid at the 2-position of phosphatidyl choline and suggested that this provided a suitable method for modifying preformed phospholipid molecules and adapting their fatty acid composition to the requirements of a particular tissue. Studies with phospholipase A have proved that in many animal tissues, saturated fatty acids are esterified mainly in the 1-position while unsaturated fatty acids are esterified mainly at the 2-position. (This is not a universal rule as some bacterial lipids show quite the opposite pattern). It is also known that the various phospholipids have fatty acid patterns different from each other and from neutral lipids. These differences could arise from the reaction described by Lands (called 'transacylation') or another possibility is that the fatty acid composition of each phospholipid is inherent in its diglyceride and that (e.g.) the enzyme *CDP-choline-diglyceride transferase* 'selects' diglyceride molecules (originating from phosphatidic acid) of appropriate composition. Current research in this area is directed mainly towards answering this question. The experiments are not clear cut and the evidence suggests that the selection occurs at both enzyme levels, although how the enzymes are able to make a selection is still a mystery.

The biosynthesis of some complex amino acid- and sugar-containing phospholipids of bacteria.

Most of the early work which unravelled the biosynthetic pathways leading to phospholipids used enzyme preparations from animal tissues. There is a large gap in our knowledge of plant phospholipid biosynthesis but more recently many scientists have been turning to the bacteria where, with more complex lipids to study, some fascinating new biosynthetic mechanisms have been revealed.

Phospholipids, as we shall show in a later chapter, are usually associated in tissues with proteins as a *lipoprotein complex*. When the lipids are extracted with solvents, the lipid-protein bonds are broken but the physical properties of phospholipids are such, (they have a long chain fatty acid *lipid end* and a *polar end* which give them a dual character) that the lipids draw many *polar* or water-soluble substances into solution with them. These substances include free amino acids and at first many claims were made for the existence of complex lipids containing amino acids. You may come across many terms in the literature — *lipo-amino acids, lipopeptides, proteolipids* — which are better avoided because they give little or no indication of whether or not there is a covalent bond between lipid and non-lipid components. The best characterized compounds that *do* have a covalent linkage are the amino acid esters of phosphatidyl glycerol which have so far been found exclusively in bacteria. The first of these were discovered and characterized by the British biochemist Marjorie Macfarlane, working at the Lister Institute in London.

These were the O-L-lysyl ester of phosphatidyl glycerol found in *Staphylococcus aureus*, and the O-L-alanyl ester in *Clostridium perfringens*. Subsequently, lysine, alanine, ornithine and glycine esters have been found in a number of other organisms. An interesting feature of these compounds is that their concentration in the cell is strongly dependent on the environment of the bacteria. One of the outstanding features about the bacteria which makes them such useful tools in biochemical research and distinguishes them from the tissues of higher animals is that their environment can be changed at will by the experimenter and the effects which this produces can be traced in the compounds present in the membrane or in the cell interior. The phospholipids are present almost exclusively in the membrane of the bacteria and hence they can act as an 'indicator' of the state of the membrane. The amino acyl phosphatidyl glycerols were found to accumulate when the bacteria were growing in an acid medium (often due to the lactic acid produced by the fermentation processes of the bacteria themselves). In media where the pH remained at 7 or had been artifically maintained at 7, the proportion of amino acyl phosphatidyl glycerols was small, whereas the parent phospholipid (phosphatidyl glycerol) accumulated.

These observations, while throwing light on the metabolic state of phosphatidyl glycerol derivatives, tell nothing about the pathways for their biosynthesis. The American biochemist, W.J. Lennarz, looked for a cell-free preparation from *S. aureus* which might

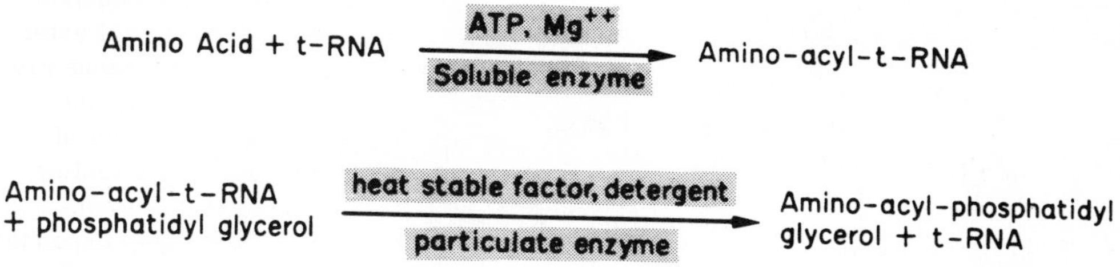

$$\text{Amino Acid} + \text{t-RNA} \xrightarrow[\text{Soluble enzyme}]{\text{ATP, Mg}^{++}} \text{Amino-acyl-t-RNA}$$

$$\text{Amino-acyl-t-RNA} + \text{phosphatidyl glycerol} \xrightarrow[\text{particulate enzyme}]{\text{heat stable factor, detergent}} \text{Amino-acyl-phosphatidyl glycerol} + \text{t-RNA}$$

Fig. 4.8. Formation of amino acyl phosphatidyl glycerols.

be capable of synthesizing amino acyl phosphatidyl glycerols from U-^{14}C-L-lysine. A recombination of three fractions from the disrupted bacterial cells was found to be necessary — a 'particle', (or readily sedimentable) fraction, a 'soluble' enzyme and a heat stable factor obtained after boiling the soluble fraction of the cells. In the presence of ATP, Mg^{++}, and a detergent (such as the sodium salt of a fatty acid), U-^{14}C-L-lysine was incorporated into a lipid which could be identified as lysyl phosphatidyl glycerol. The real clue to the *activation* of the amino acid came when Lennarz tried to discover what would be the effect of certain inhibitors on the reaction. The enzyme ribonuclease, which degrades RNA into its constituent nucleotides, completely prevented the reaction from occurring. The Baltimore Research Group concluded that probably lysyl-*t*-RNA was being formed in the reaction. This is the same complex between an amino acid and *transfer RNA* (*t*-RNA) which carries specific amino acids to the ribosomes in protein biosynthesis. They were soon able to show that the need for the 'soluble' enzyme could be eliminated by using ^{14}C-lysyl-*t*-RNA as the substrate. In this way they demonstrated a novel reaction in which a mechanism, already well known in protein biosynthesis, was employed in lipid metabolism (Fig. 4.8).

A biosynthetic pathway which has not yet been worked out (at the time of preparing this textbook!) but which should prove to be very fascinating, is that leading to the formation of glucosaminyl phosphatidyl glycerol (see Table 4.1). Other sugar-containing phospholipids, namely the phosphatidyl inositol mannosides of *Mycobacterium phlei* are formed by stepwise addition of the mannose molecules to phosphatidyl inositol. Following the usual pattern in carbohydrate chemistry (indeed in biosynthesis generally) the sugar molecules are transferred from an 'energy rich' molecule, in this case, GDP-mannose, in a so-called *group transfer reaction*.

Most tissues contain phospholipid-degrading enzymes or 'phospholipases'

The idea which has been emphasized

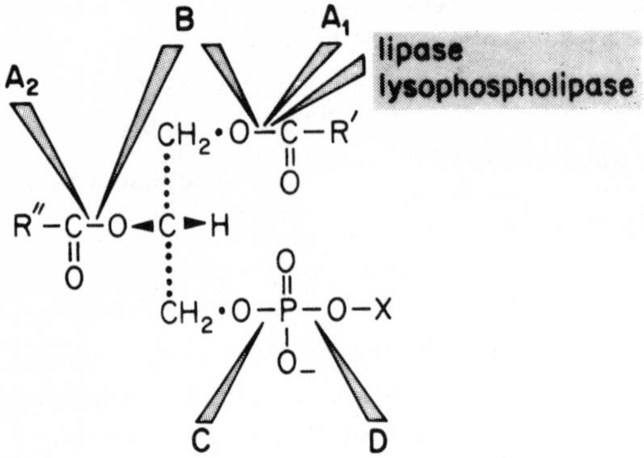

Fig. 4.9. Site of action of phospholipase A,B,C and D.

repeatedly in this chapter, that different parts of the phospholipid molecule are being replaced independently, implies the presence in the cell of enzymes which can hydrolyse the different bonds in the molecule. Over the years, enzymes have been identified which hydrolyse at each of the fatty ester bonds, the phosphorus — glycerol bond and the phosphorus — base bond (Fig. 4.9). At times it seems as if each investigator has added a new name and a new enzyme so that the nomenclature has been confusing, to say the least. Use of the recently introduced Enzyme Commission Nomenclature (see the reference section at the end of chapter 1) will do a lot to restore sanity, but for shortness and convenience most authors still use the ABCD nomenclature.

The specificity of phospholipase A

The first confusion arose due to a mis-understanding of the specificity of the enzyme from certain snake venoms (3.1.1.4 *phosphatide acyl hydrolase, phospholipase A*) which releases one mole of fatty acid from a diacyl phospholipid. The original studies showed quite clearly that unsaturated fatty acids were released from egg or liver phosphatidyl cholines and so it was quite natural to assume that the enzyme was specific for unsaturated acids. When fully saturated or fully unsaturated lipids were the substrates, however, *one mole* of fatty acid was still released from one mole of phospholipid and it was realised that the enzyme must be specific for one or other of the two positions. Early chemical analyses of the products of phospholipase A hydrolysis seemed to indicate that the enzyme released the fatty acid from the *1-position*. When it was discovered that the enzyme also liberates fatty acid from *plasmalogen* (where it is known quite definitely to be located at the 2-position) the problem was reinvestigated more rigorously.

The final, unequivocal proof of the 2-position specificity of this enzyme demonstrates one of the ways in which first-class organic chemists can advance knowledge of biosynthetic mechanisms in the lipid field, just as the work of chemists such as the American H.G. Khorana and the Israeli, E. Katchalski have done for nucleic acid and protein biosynthesis respectively. The Dutch chemists G. de Haas and L. van Deenen prepared 'mixed acid' phosphatidyl cholines and ethanolamines by elegant new synthetic techniques so that the 1- and 2-position fatty acids were specified exactly. Hydrolysis of these lipids with phospholipase A could then demonstrate a 2-position specificity directly. Studies with synthetic analogues of the substrate were able to pinpoint the precise structrual and stereochemical features necessary for the enzyme to hydrolyse a fatty ester bond. The lipids of the 2,3-diacyl-1-sn-glyceryl phosphoryl-X series were not hydrolysed, but from 1,3-diacyl-2-sn-glycerophospholipids only the 1-position fatty acid was hydrolysed. The enzyme seems to require the presence of only one fatty ester linkage adjacent to the alcohol-phosphate bond and the carbon atom to which this fatty acid is attached must have a precise stereochemical configuration.

Although the main source of this enzyme has been snake venom, the enzyme (now universally named phospholipase A_2) has a very wide distribution. It is generally believed that the production of small amounts of lysophospholipids by tissue phospholipase A's, which are then available for reacylation by the 'Lands mechanism', is one of the chief ways by which the fatty acid composition of the tissue lipids can be 'tailored' to suit the requirements of the cell. Hence, phospholipase A becomes a key enzyme in the regulation of lipid composition. Research on the specificity of the enzyme has now gone full circle by the discovery of another enzyme, present in a microsomal fraction obtained from rat liver, called phospholipase A_1). This enzyme catalyses the release of the fatty acid from the 1-position and makes for greater versatility in the regulatory process. Confusion still exists, however, because the triglyceride lipase from pancreas is also known to release the 1-position fatty acid from phospholipids as well as from glycerides and it could be that in some cases these two enzymes are being confused.

Studies on the 'active centre' of phospholipase A are just in sight.

Very few enzymes of lipid metabolism have had even their primary structure (i.e. the sequence of constituent amino acids in the chain) elucidated. The phospholipase A_2 from pig pancreas is an exception to this rule and in the last few years, de Haas and the Dutch School have purified the enzyme and worked out its amino acid sequence. The activity of the freshly isolated enzyme is rather low but upon 'ageing', the activity increases quite dramatically in contrast with most enzymes which the biochemist has to deal with. De Haas discovered that this 'activation' is due to cleavage of a small peptide from the protein chain by a proteolytic enzyme.

Neither the peptide nor the complete molecule has phospholipase activity but the smaller protein does. This phenomenon occurs in several proteolytic enzymes as well, for example in the conversion of chymotrypsinogen into chymotrypsin, and is presumably a mechanism by which some pancreatic phospholipids are protected from continual degradation.

The substrates for phospho-lipases are large aggregates of lipids with a critical electrical charge on the surface.

~~Various~~ other fatty acid releasing enzymes have been described but they are as yet ill defined. The existence of a 'phospho-lipase B', which is reputed to hydrolyse both fatty ester linkages, is controversial. It is possible that this consists of a mixture of phospholipase A and lyso-phospholipase activities. The latter enzyme (EC.3.1.1.5, *lysolecithin acyl hydrolase, lysophospholipase*) removes the 1-fatty acid from a 1-monoacyl glycerophospholipid. Whatever the true picture may be, an interesting problem of lipid enzymology does arise from a study of 'phospholipase B-type activity'. The enzyme from the mould *Penicillium notatum* will not attack pure phosphatidyl choline but is active in the presence of 'activators' such as phosphatidyl inositol or other lipids which give the phosphatidyl choline *micelles* a net negative charge. In most other areas of enzymology the biochemist has to study the reaction between the enzyme and a water-soluble substrate where he is pre-sumably dealing with single substrate molecules. When a lipid is the substrate, the molecules exist singly in true

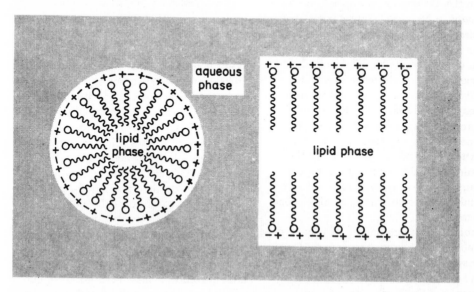

Fig. 4.10. Phospholipid micelles.

solution below a certain concentration called the *critical micelle concentration* (CMC). Above this concentration phospholipids, which have a *polar* ('phosphoryl-X') end and a *non-polar* (fatty acid) end, tend to form aggregates or micelles which are arranged with the polar ends in the aqueous environment. (See chapter 6 for fuller explanation). Two such possible arrangements are illustrated in Fig. 4.10. It is with these micelles that the enzyme must interact. The net charge distribution on the micellar surface will depend on the composition of the polar end groups and if the net charge at the active site of the enzyme is repulsive toward the micelles, then no enzyme-substrate interaction can take place. Similar physicochemical arguments explain why other phospholipases such as A, C and D are activated by the presence of diethyl ether in the incubation mixture. Either the solvent allows more ready access of the hydrocarbon chains to the enzyme, reduces micellar size, or helps to remove accumulated fatty acids. These kinds of experiments are somewhat artificial, as the real substrates attacked by the enzymes in the living cell are bound as lipoproteins. Indeed there is convincing evidence that protein-bound lipids are hydrolysed *in vitro* more rapidly than extracted purified lipids.

In accordance with the facts about independent turnover of phosphate and 'base' moieties, other enzymes, named phospholipases C (EC.3.1.4.3, *phosphatidyl choline: choline phosphohydrolase*) and D (EC.3.1.4.4, *phosphatidyl choline phosphatide hydrolase*) are known to hydrolyse each of the two phosphodiester links (see Fig. 4.9). The most important sources of these enzymes are *Bacillus cereus* and *Clostridium perfringens* in the case of phospholipase C, and cabbage leaves in the case of phospholipase D. Although phospholipase C activity has been detected in some animal tissues, phospholipase D has been found only in plants. As well as catalysing the splitting of group 'X', it has been found to catalyse the exchange of group 'X' with various other alcohols or bases. This reaction is known as *transphosphatidylation* and was first noticed because phosphatidyl methanol was formed when plant lipids were extracted with methanol. The enzyme may therefore be important not only in the breakdown of phospholipids but in lipid transformation by 'base exchange'. In addition there are also enzymes for dephosphorylating the polyphosphoinositides, cleaving the vinyl ether linkage of plasmalogens and the N-acyl bond of sphingomyelin.

The hydrolytic enzyme phosphatidate phosphohydrolase is more important in synthesis than in degradation of phospholipids.

The key enzyme in the whole series which degrade phospholipids is phosphatidate phosphohydrolase, the enzyme which splits off the phosphate from phosphatidic acid to produce diglyceride. We have already learned that it is this diglyceride which is not only the precursor of the acyl moiety of many important phospholipids but a significant precursor of triglyceride as well. The activity of this enzyme is important, therefore, not in the catabolism of phospholipids but in regulating their synthesis. As might be expected, its distribution throughout species and tissues is very extensive and its intracellular distribution is also very wide. G. Hübscher and his colleagues in Birmingham, England have purified the enzyme and studied it carefully. The enzymes in the mitochondrial and microsomal fractions are firmly bound to the membranes of those fractions but can be extracted by partial degradation with ribonuclease followed by treatment with n-butanol. Interestingly, the very tightly bound mitochondrial or microsomal enzymes can easily hydrolyse a dispersion of phosphatidic acid in water (this is the usual way of assaying the enzyme) whereas the form of the enzyme found in the soluble fraction of the cell is ineffective on this substrate. Conversely, membrane-bound phosphatidate is not degraded by membrane-bound hydrolase, but extremely rapidly by 'soluble' hydrolase. It is this soluble enzyme which so powerfully stimulates glyceride synthesis as we saw in chapter 3.

Improved analytical techniques have been the key to our increased knowledge of lipid metabolism

The recent advances in our knowledge of the pathways of synthesis and breakdown of phospholipids have followed in the wake of greatly improved methods for their separation and analysis. We shall devote the remainder of this chapter to a description of some of the more modern or more widely used of these techniques.

We have given a general account of the procedures for extraction, washing, and purifying lipids in Chapter 1 and we would encourage the student to re-read parts of that chapter at this stage. The most commonly used solvent for extracting phospholipids from tissues is chloroform-methanol (2:1, v/v). In some cases, notably in the extraction of poly-phosphoinositides from brain, preliminary extraction with acetone is desirable for reasons not entirely understood, and for complete extraction of this class, in-clusion of a small amount of HC1 in the chloroform — methanol is sometimes necessary. The *amphipathic* character (presence of hydrophilic and hydrophobic regions) of phospholipids inevitably leads to 'cosolubilization' of small water-soluble molecules even though the solvent is essentially non-polar. Such contaminant have to be removed, usually by washing with a salt solution such as 1% sodium chloride, although, here again, care must be taken not to lose those highly polar phospholipids such as phosphatidic acid or the inosities which are appreciably water-soluble and either partition into the water phase or accumulate at the 'inter-face'. A more modern development for removing water-soluble contaminants is to pass the lipid solution over some kind of 'molecular sieve' such as a specially designed 'Sephadex' which allows the lipid molecules to pass through the gel, retaining smaller molecules.

Before attempting a separation, it is useful and often essential, to know the quantity of material available. Weighing oily compounds is often difficult and in any case inaccurate for small quantities and an estimation of phosphorus content is often the best method of quantitation as long as water-solubles have been removed. The usual procedure depends on measurement of the blue colour formed when the complex between phosphate ion and molybdate is reduced.

Silicic acid has proved the most versatile chromatographic support material.

Use of silicic acid as a stationary phase in phospholipid chromatography has a long history. Its application in general lipid separations has already been described in earlier chapters but in recent years, new ways of using the material have revolutionized phospholipid separations. For large quantities, a column of silic acid ('activated' by heat to remove water, and sieved to remove smaller particles and improve flow rates) is still the best. The difficulty here at present is the detection of material eluted from the column. One must collect 'fractions' and make a determination on each fraction, such as a phosphorus analysis, radioactivity detection or secondary run on TLC (see later), in order to plot the eluted peaks. A 'continuous' monitoring system based on some physical property of the eluted material is being currently developed but its sensitivity and general applicability are not yet comparable with the GLC system. Historically, the next development in the use of silic acid was paper chromatography, for which the papers had been impregnated with the support material. This method can be very useful in certain circumstances but for

general purposes it has been almost entirely superseded by thin layer chromatography (TLC).

Quite good separations of most phospholipids can be achieved on thin layers (0·25 mm) of silica gel 'activated' by heating at about 100° for an hour just before use. The critical point about most of the solvent systems for separating phospholipids by TLC is the proportion of water. For general purposes, a mixture of chloroform, methanol and water in the proportions 65:25:4 is found to be most useful although we should emphasize that, in the first place, this solvent will rarely separate *every* phospholipid in a complex mixture and second, that as in all types of chromatography used for analytical purposes the experimenter should not be satisfied with identification in *one* solvent system alone. The author of this chapter is addicted to the use of 'microscope slide' TLC plates. These are very useful for such things as quick checks on fractions eluted from a large preparative column or for following the course of a chemical reaction. They have the advantage that they need no activation— they are made by quickly dipping a slide into a slurry of silica gel in 1:1 by volume of chloroform-methanol mixture (the correct consistency is found by experience) and allowing to dry. They can be used immediately and when developed in a small jar, take only five or ten minutes to run, depending on the solvent.

For preparative purposes, too, chromatography on glass plates can be as useful and more convenient than column chromatography by the simple expedient of using a thick layer (up to 1 mm). An example of their use might be the separation of *lyso* compound and fatty acid after an incubation of a phospholipid with phospholipase A. By scraping off the areas containing the products and methylating the material *directly on the silica*, the fatty acid methyl esters can be formed, analysed by GLC, and their distribution on the 1- and 2-postions of the lipid established.

Identification of compounds on a TLC plate is important. Most of the colour reactions used in paper chromatography are applicable (e.g. 'molybdate blue' reaction for phosphate, ninhydrin for free amino groups, periodate-Schiff reagent for vicinal hydroxyls such as are present in phosphatidyl glycerol). A general, non-discriminating method which can be used, is charring with sulphuric acid. This is especially suitable if the plate is to be photographed. Even with this non-specific reagent, different classes of lipids give characteristic colours in the early stages of charring if the plate is heated carefully. All these methods are of course destructive and cannot be used if the compounds are to be recovered. In this case a reagent which fluoresces, such as Rhodamine, is usually chosen, and the plate is examined under ultraviolet light. When the lipids are eluted from the plate with chlorofrom-methanol mixtures, this reagent remains adsorbed on the silica. Brief exposure to iodine vapour can be useful, and even light spraying with water if there is plenty of lipid material on the plate.

Recently, major advances have been made by the use of thin layers of silica impregnated with 10% silver nitrate solution. Plates of this type must be stored in the dark in a constant humidity tank. These plates are used for separating the members of a single phospholipid *class* (e.g. phosphatidyl cholines) into 'species' according to the total number of double bonds in the molecule. Thus a molecule containing two saturated fatty acids runs ahead of one containing one saturated and one mono-unsaturated acid, and so on. In this way, some intriguing facts have come to light, such as, for example, that some species have a greater phosphate 'turnover' than others. When only the fatty acids and not the phosphate moieties are to be investigated, many people prefer to first hydrolyse the phospholipid with phospholipase C to produce the corresponding diglycerides. *Argentation* TLC can then give a better separation of diglyceride species than of the intact parent phospholipids. Phospholipid species with fatty acids of the same degree of unsaturation, but different chain lengths can be separated by *reversed phase* chromatography.

Materials other than silica can of course be used. Alumina has been useful for separations of rather simple mixtures such as egg yolk phospholipids. Separation must be rapid, owing to the rather basic nature of the stationary phase. Diethylaminoethyl (DEAE)-cellulose, too, can be used in a similar way to silicic acid and has the advantage of very much faster running, and possibilities of separating pairs of compounds not well separated on silicic acid. Nevertheless, it must be emphasized that each method needs to be used in conjuction with some other method.

Nowadays it is usual to begin with some kind of column method, followed by chromatography of the material from column 'peaks' on thin layers.

An alternative to separation of intact lipids is to remove fatty acids by mild hydrolysis leaving a mixture of phosphodiesters which can be separated on paper by chromatography or, more usefully, by electrophoresis. This allows separation of classes not easily separated on silicic acid (e.g. phosphatidyl ethanolamine and phosphatidyl serine), and is useful for identification purposes. On a large scale, the phosphodiesters can be separated on columns of ion exchange resin. The method will naturally not distinguish between diacyl and monoacyl phospho-lipids and plasmalogen analogues must be broken down by a further step involving acid hydrolysis. Strong acid hydrolysis of phospholipids can release the group 'X' (choline, ethanolamine, inositol, etc.), glycerol phosphate, inorganic phosphate, glycerol and fatty acids which can then all be identified by standard analytical techniques.

In this chapter, we have described the structural features of phospholipids, how they are built up and broken down within the cell, and how they may be extracted, separated, purified and identified. We have learned little about where they are located in the cell and what their function may be. We will consider these and related questions in a later chapter.

SUMMARY

Phospholipids are mixed esters of fatty acids and phosphoric acid with the alcohols glycerol or sphingosine. A wide spectrum of chemical species is made possible, first, by very considerable variation of the types and combination of fatty acids and second, by esterification of different organic bases, amino acids and alcohols to the phosphate group. They derive their lipid properties from the long chain fatty acid moieties but also have a considerable polar character donated by the ionization of the phosphate and base groups. In physical properties they 'bridge the gap' between the completely water insoluble *neutral lipids* and molecules which form true aqueous solutions.

Phospholipids are ubiquitous in animal and plant cells and are often undergoing continuous breakdown and resynthesis. Biosynthesis of complete phospholipid molecules (*de novo*) occurs by two main types of pathway, one involving transfer of 'phosphorus-base' from a water soluble nucleotide, *cytidine diphosphobase*, to a di-glyceride; the other involving transfer of phosphatidic acid from a lipid soluble nucleotide, *cytidine diphosphodiglyceride* to the base. Alternatively, completed molecules may be altered to suit the cell's requirements by 'exchange reactions' involving cleavage of the fatty acids and reacylation, or cleavage of the base and re-esterification. Glycerophosphate, arising mainly from the glycolysis pathway, forms the building block for the parent molecule, phosphatidic acid, by esterification with acyl-CoA.

Widely distributed enzymes are capable of catalysing the hydrolysis of either one or both fatty acids, the glycerol-phosphate linkage or the phosphate-base linkage. The key hydrolytic enzyme, phosphatidate phosphohydrolase, is strangely enough, more important in biosynthesis than catabolism as it releases inorganic phosphate from phosphatidic acid to produce the diglyceride, which is the lipid precursor for many complex phospholipids and for triglyceride biosynthesis.

Combinations of modern chromatographic techniques such as silicic acid or DEAE-cellulose column chromatography and TLC are capable of separating the components of most natural phospholipid mixtures. Silica layers impregnated with silver nitrate can separate species of phospholipids differing only in the number of double bonds contained in their fatty acids. This technique has brought to light the remarkable fact that phospholipids of the same 'class' but differing only in fatty acid composition may have vastly different 'turnover' rates and even different metabolic fates.

BIBLIOGRAPHY

1. KENNEDY E.P. (1961). Biosynthesis of complex lipids *Fed. Proc.* **20**, 934. An early review, when the basic pathways had just been worked out.

2. WITTCOFF H. (1951). The Phosphatides. Reinhold, New York. Good reading for those interested in the foundations of the subject but probably only available in a well established library.

3. ANSELL G.B. and HAWTHORNE J.N. (1964). Phospholipids. Elsevier, Amsterdam. A comprehensive review of all aspects of animal phospholipids.

4. DAWSON R.M.C. (1966). Metabolism of animal phospholipids and their turnover in cell membranes. *Essays in Biochemistry* **2**, 69, edited by P.N. Campbell and G.D. Greville (The Biochemical Society), Academic Press, London and New York.

5. VAN DEENEN L.L.M. and DE HAAS G.H. (1966). Phosphoglycerides and Phospholipases. *Ann. Rev. Biochem.* **35**, 157.

6. LENNARZ W.J. (1966). Lipid metabolism in the bacteria. *Advances in Lipid Research* **4**, 175.

7. GOLDFINE H. (1968). Lipid chemistry and metabolism. *Ann. Rev. Biochem.* **37**, 303. The most up to date word on the subject.

8. JAMES A.T. and MORRIS L.J. (eds.), (1964). New biochemical separations. D. van Nostrand, London. There are several articles in this collection dealing with various aspects of lipid chromatography.

9. MARINETTI G.V. (1967). *Lipid Chromatographic Analysis* Vol. 1, Marcel Dekker, New York.

5 Glycolipids and Sulpholipids

Glycolipids are compounds which have the solubility properties of a lipid and contain one or more molecules of a sugar. Bacteria have high molecular weight polymers containing lipids and sugars, called lipopolysaccharides, which are quite distinct from glycolipids in being water soluble; these will be described in Chapter 6. The word *sulpholipid* is used here to describe any lipid containing sulphur, although the term is often used in a more restricted sense to describe a certain class of sulphur-containing lipids, as we shall see later. It is important to realize that the terms glycolipid and sulpholipid do not refer to any chemically distinct groups of compounds, but cover a wide range of chemical types; they merely indicate a particular chemical feature of the molecule i.e. the presence of sugar or sulphur. Sulpholipids are often glycolipids and there is also an overlap between glycolipids and phospholipids. Several phospholipids which contain sugars have been described in the last chapter, and will not therefore be discussed again: [glucosaminyl phosphatidyl glycerol p. 128; phytoglycolipid, p. 130; phosphatidyl glucose, p. 128; mannosyl phosphoinositides, p. 128].

Classification is difficult because of this overlapping and the nomenclature is confusing because different authors have often applied their own systems. To draw an analogy with the phospholipids, we will choose to divide glycolipids into two broad classes — those based on the long chain fatty base, *sphingosine* and those based on *glycerol*. There are other methods of course, and the student should be aware of this when reading other textbooks and articles. This method leaves quite a large heterogeneous group which does not fall into either class.

GLYCOSYL CERAMIDES

*The long chain base sphingosine
or a related base is linked to the
acyl chain through an amide bond
and to the sugar moiety through
a glycosidic bond.*

In both phospholipids and glycolipids
the long chain base, *sphingosine*,
[*D-erythro*-2-amino-*trans*-4-octadecene-
1,3-diol; 4D-Sphingenine] almost always
has the fatty acid linked as an amide
rather than an O-ester. N-acyl-sphingosine is called *ceramide* and this forms
the common feature of the first class of
glycolipids: glycosyl ceramides.

Monoglycosyl ceramides. The simplest
glycosphingolipids of most mammalian
tissues are the monoglycosyl ceramides
or *cerebrosides*. The deacylated product
of galactosyl ceramide, O-sphingosyl
galactoside is called *psychosine*
(Fig. 5.1 and Table 5.1). Variations are
possible in all three moieties: the base,
fatty acid and sugar, as indicated in

Table 5.1. More complex glycosphingolipids such as cerebroside sulphates and
ceramide oligosaccharides are also
described in Table 5.1.

Gangliosides. This class of glycosphingolipids contains one or more molecules of
sialic acid linked to one or more of the
sugar residues of a ceramide
oligosaccharide. Sialic acid is N-acetyl
neuraminic acid (NANA; Fig. 5.3).
Glycolipids with one molecule of sialic
acid are called monosialogangliosides;
those with more sialic residues are
di- and trisialogangliosides.

Disialogangliosides may have each
sialic moiety linked to a separate sugar
residue or both sialic acids may be
linked to each other and one of them
linked to a central sugar residue. Some
common gangliosides are shown in
Table 5.2. Because of their complex
structure and hence cumbersome
chemical names, many shorthand notations

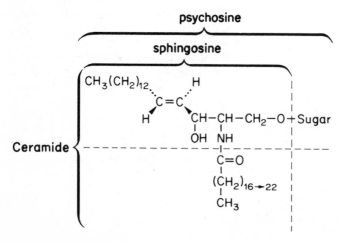

Fig. 5.1. Basic structure of
glycosyl ceramides.

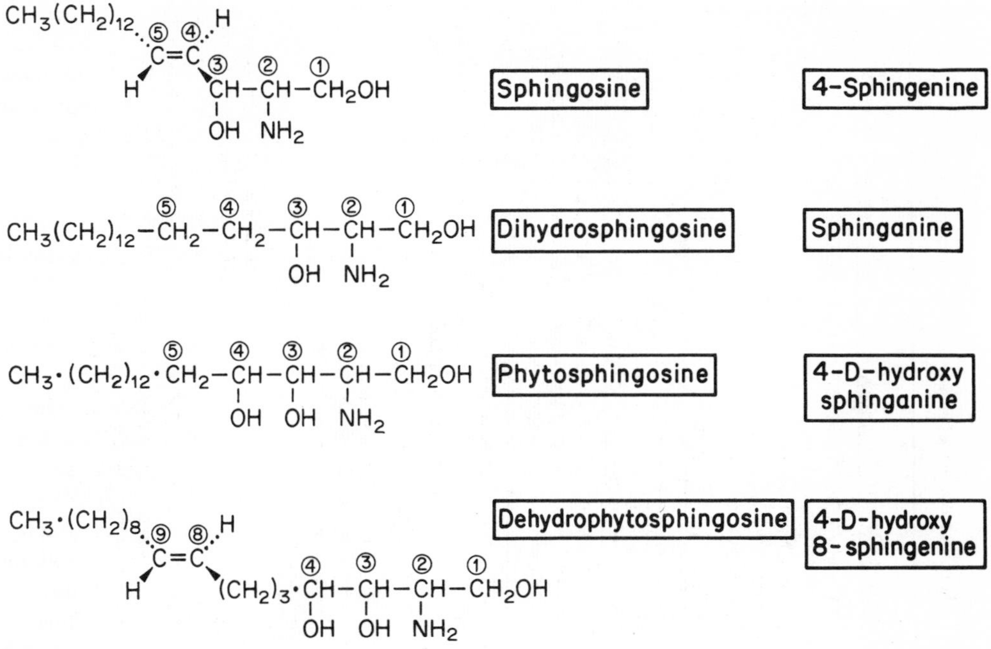

Fig. 5.2. Variations of the sphingosine base.

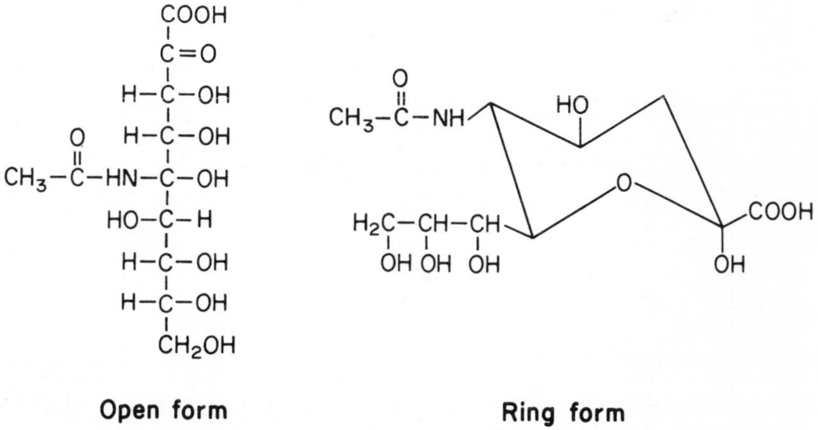

Open form   **Ring form**

Fig. 5.3. N-acetyl-neuraminic acid (NANA) or sialic acid.

Table 5.1. The Structure of Some Glycosyl Ceramides
[Cer = ceramide; glc = glucose; gal = galactose; GalNAc = N-acetyl galactosamine; see Fig. 5.1]

Accepted nomenclature	Old trivial name	Structure	Description	General remarks
MONOGLYCOSYL CERAMIDES	Cerebroside	Cer-gal Cer-glc In all glycosyl ceramides and gangliosides there is an O-glycosidic linkage between the primary hydroxyl of sphingosine and C-1 of the sugar	'Cerebroside' originally used for galactosyl ceramide of brain but now widely used for monoglycosyl ceramides. Sugar composition depends largely on tissue. Brain cerebroside mainly galactoside while serum has mainly glucose. In animals the highest concentration is in brain. Monogalactosyl ceramide is largest single component of myelin sheath of nerve. Intermediate concentrations in lung and kidney. Also found in liver, spleen, serum with trace amounts in almost all tissues examined.	*Fatty Acids*: glycosphingolipids are characterized by having large concentrations of (a) hydroxy acids (b) long chain odd and even fatty acids, in comparison with other lipids. The hydroxy acids include α-hydroxy acids formed by α-oxidation. This oxidation mechanism is probably also responsible for odd chain acids. Typical acids are: 22:0 (behenic) 24:0 (lignoceric); 24:1 (nervonic); α-OH-24:0 (cerebronic).
Sulphatide	Cerebroside sulphate	cer-gal-3-sulphate	Very generally distributed like cerebrosides. Fatty acid and base composition similar.	
DIGLYCOSYL CERAMIDES (a) Lactosyl Ceramide	cytosides Cytolipin H	Cer-glc(4←1)-gal	Major diglycosyl ceramide. Widely distributed. Key substance in glycosyl ceramide metabolism. It may accumulate, but as it is a precursor for both ceramide oligosaccharides or gangliosides it may be present in trace quantities only	*Base*: C_{18}-sphingosine is dominant. Smaller amounts of C_{18}-dihydrosphingosine (esp. leaves, wheat flour)C_{18}-phyto- and dehydrophytosphingosine also occur (see also Fig. 5.2).

Table 5.1. Cont.

Accepted nomenclature	Old trivial name	Structure	Description	General remarks
(b) Digalactosyl ceramide		Cer-gal(4←1)-gal	Minor diglycosyl ceramide. Found esp. in Kidney (human and mouse)	In general, each *organ* has a dominant type of glycolipid but its nature may depend also on the *species*, e.g. mono-glycosyl ceramides are dominant in brain; tri-hexosides and amino-glycolipids in red cells. Modern analytical tech-niques are revealing that most types are widespread among tissues. Cere-brosides are widespread in higher plants, the best characterized being those of bean leaves and wheat flour. Glucose is probably the only sugar. Sphingolipids are rare in micro-organisms.
TRIGLYCOSYL CERAMIDES				
digalactosyl glucosyl ceramide		cer-glc(4←1)gal (4←1)gal	Source: kidney, lung, spleen, liver. Most analyses have been on human tissue.	
TETRAGLYCOSYL CERAMIDES				
(a)	aminoglycolipid globoside	Cer-glc(4←1)gal (4←1)gal(3←1)- -β-N-acetyl galactosamine	Most abundant lipid in human erythrocyte stroma	
(b)		Cer-glc(4←1)gal-(4←1) gal (3←1)- α-N-acetyl galactosamine	So-called 'Forsmann Antigen'	
(c)	Asialoganglioside	Cer-glc(4←1)gal-(4←1) galN Ac-(3←1) gal	Basic ganglioside structure without N-acetyl neuraminic acid (sialic acid) residues. Intermediate in ganglioside biosynthesis (rat, frog, brain)	

153

Table 5.2 **Structure of Some Gangliosides**

Class	Structure	Wiegandt shorthand	Svennerholm shorthand	Composition and occurrence
MONOSIALO-GANGLIOSIDE	(a) Cer-glc(4←1)gal(4←1)galNAc(3←1)gal β linkages; (3←2) NANA	G_{GNT} 1	G_{M1}	*Major bases*: C_{18} and C_{20}-Sphingosines. Minor amounts of dihydroanalogues. *Fatty Acids*: Large amounts of 18:0 (86–95% in brain)
Tay-Sachs ganglioside	(b) Cer-glc(4←1)gal(4←1)galNAc β linkages; (3←2) NANA	G_{GNT-II} 1	G_{M2}	*Occurrence*: Mainly in gray matter of brain but also in spleen, erythrocytes, liver, kidney. Modern analytical techniques have shown them to be present in a much wider range of tissues than previously realized. Main gangliosides of human brain are G_{GNT} 1, 2a, 2b, 3a.
Haematoside	(c) Cer-glc(4←1)gal(3←2)NANA β linkages	G_{Lact} 1	G_{M3}	Gangliosides appear to be confined to the animal kingdom. In man, cattle, horse, main ganglioside outside brain is G_{Lact} 1. N-glycolyl-neuraminic acid is chief sialic acid in erythrocyte and spleen ganglioside of horse and cattle.
DISIALO-GANGLIOSIDE	(a) Cer-glc(4←1)gal(4←1)galNAc(3←1)gal β linkages; (3←2) NANA, NANA	G_{GNT} 2a	G_{D1a}	*Physical properties*: Insoluble in non-polar solvents; form micelles in aqueous solution.
	(b) Cer-glc(4←1)gal(4←1)galNAc(3←1)gal β linkages; (3←2) NANA, NANA(8←2)NANA	G_{GNT} 2b	G_{D1b}	
TRISIALO-GANGLIOSIDE	Cer-glc(4←1)gal(4←1)galNAc(3←1)gal β linkages; (3←2) NANA, NANA(8←2)NANA NANA	G_{GNT} 3a	G_{T1}	

Abbreviations in column 2 are the same as in Table 5.1. NANA = N-Acetylneuraminic acid (sialic acid). Wiegandt Abbreviations:— G = ganglioside. Subscript denotes sialic-free oligosaccharide. Arabic numeral gives number of sialic residues. T = basic tetraose. Tr = triose; Lact = lactose. Trisaccharides and disaccharides that can be derived from the basic tetraose are distinguished by I, II, etc., I describing the sugar that originated from the non-reducing end of the tetrasaccharide. a, b refer to positional isomers with respect to sialic residues.

have been employed. One of those most commonly used is that introduced by the Swedish scientist, L. Svennerholm. The parent molecule is denoted as G_{M1} and other derived structures are shown in Table 5.2. A more recent system, due to the German biochemist, H. Wiegandt, has also been included in Table 5.2 because it may well be the general method of choice before very long. It has the advantage that once the symbols have been learnt, the structure can be worked out from the shorthand notation, and it can be applied to non-sialic acid containing glycolipids as well. In the following text, gangliosides will be denoted by the Wiegandt notation followed by the Svennerholm notation.

Research on the biosynthesis of glycosphingolipids has lagged behind studies on their breakdown because of the importance of the catabolic pathways in diseases of lipid metabolism.

Whereas the main pathways of phospholipid biosynthesis in a whole range of tissues and organisms are known in great detail, those for glycolipid biosynthesis are in most cases uncertain. Much of the work on glycosphingolipid metabolism has been done from the point of view of studying the diseases associated with these lipids, known collectively as *lipidoses*. In most cases these disorders arise because the tissue lacks a key enzyme involved in the *breakdown* of the lipids and so the substances accumulate to an abnormal degree. It is not surprising, therefore, that the catabolism of glycosphingolipids has received more attention than their biosynthesis.

As in all branches of lipid biochemistry, the early studies of glycolipid biosynthesis were done by injecting probable radioactive precursors (sugars, fatty acids or long chain bases) into the animal and subsequently, by suitable analytical techniques, locating the different compounds into which radioactivity had been incorporated. By measuring the *specific radioactivity* of each labelled compound (the radioactivity per unit mass), the flow of radioactivity from precursor to product in a biosynthetic sequence can be observed and a good picture of the order of reactions in that sequence can be obtained. At the present time, biochemists are concentrating on studying individual steps in the pathway with cell-free extracts and partially purified enzymes.

Just as in the case of the phospholipids, we can consider the origin of the various building blocks which go to make up the whole glycolipid molecule (Fig. 5.1, 5.2).

Since 1958, the generally accepted pathway for the biosynthesis of the long chain fatty base, sphingosine, has been the condensation of palmitaldehyde (formed from palmitoyl-CoA) with the amino acid serine. The 'active form' of the amino acid is a Schiff-base complex with the coenzyme, pyridoxal phosphate, and magnesium ions. R.O. Brady and his colleagues in the U.S.A. described an enzyme in rat brain microsomal fraction which would catalyse this series of reactions:

$$CH_3(CH_2)_{14} \overset{\overset{O}{\|}}{C} \sim S \cdot CoA \longrightarrow CH_3(CH_2)_{14} CHO + CoA$$

$$CH_3(CH_2)_{14} CHO + CH_2OH \cdot \overset{-}{C} \cdot (COOH) \cdot N = \text{pyridoxal phosphate}$$

$$\xrightarrow{Mg^{++}} CH_3(CH_2)_{14} CH(OH)CH(NH_2) \cdot CH_2OH + CO_2 + \text{pyridoxal phosphate.}$$

Recently, the German biochemist, W. Stoffel, in Köln, has questioned this pathway. He has isolated the intermediate, 3-ketodihydrosphingosine, and isolated enzymes from a number of tissues which catalyse its formation by the condensation of palmitoyl-CoA and serine (without the intermediate formation of palmitaldehyde):

sphingosine. We do not yet know whether each pathway exists in different tissues, or side by side in the same tissue, or whether only one of them is correct. Sphingolipid metabolism is currently receiving a lot of attention in laboratories around the world and we may soon know the answer. The first possibility involves the addition of galactose to sphingosine

$$CH_3(CH_2)_{14} \overset{\overset{O}{\|}}{C} \sim S \cdot CoA + CH_2OH \cdot \overset{-}{C} \cdot (COOH) \cdot N = \text{pyridoxal phosphate}$$

$$\xrightarrow[\boxed{\text{brain enzyme}}]{Mg^{++}} CH_3(CH_2)_{14} \overset{\overset{O}{\|}}{C} \cdot CH(NH_2) \cdot CH_2OH$$

The next enzyme in the sequence (3-ketosphingosine reductase), reduces the keto group to a hydroxyl to yield dihydro-

by transfer from its UDP-derivative, followed by the formation of the acyl amide linkage:

$$\text{Sphingosine} + \text{UDP-Galactose} \longrightarrow \text{Psychosine} + \text{UDP}$$

$$\text{Psychosine} + \text{acyl-CoA} \longrightarrow \text{Cerebroside} + \text{CoA.}$$

sphingosine and, finally, a flavoprotein enzyme — possibly analogous to the acyl-CoA dehydrogenase of β-oxidation — converts dihydrosphingosine into sphingosine.

Cerebroside biosynthesis may proceed by acylation of sphingosine followed by addition of the sugar; or acylation of a preformed glycosyl-sphingosine.

Two pathways have been described for the formation of cerebroside from

These reactions are catalysed by enzymes in the microsomal fraction derived from rat brain and the pathways have been worked out mainly in the laboratories of E.P. Kennedy and R.O. Brady. The sphingosine isomer which takes part in this enzymic reaction is the naturally occurring *erythro* form. The mechanism for the formation of the N-acyl bond is not clear, but although it is very similar to a peptide bond, the reaction does seem to involve acyl-CoA rather than an acyl

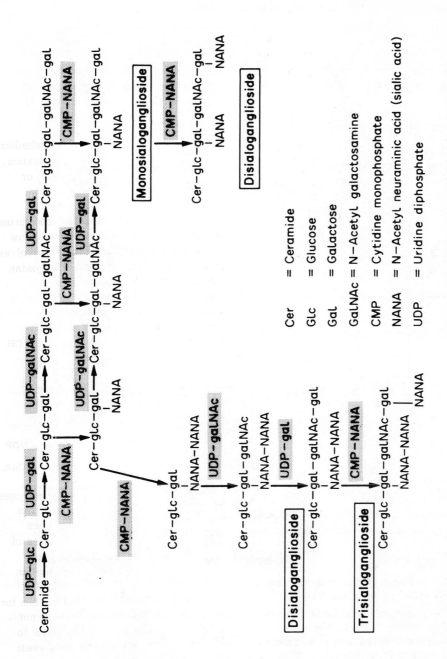

Fig. 5.4. Possible pathways of ganglioside biosynthesis.

adenylate derivative.

This pathway has been generally accepted until recently, P. Morrell and N.S. Radin proposed the following alternative:

Ceramide + UDP-Galactose $\longrightarrow$ Galactosyl-ceramide + UDP.

The enzyme, which is obtained from a microsomal fraction of mouse brain, is specific for ceramides containing hydroxy fatty acids. In this pathway, the acylation of sphingosine with acyl-CoA takes place first, followed by the transfer of the sugar from its UDP-derivative. A possible alternative route to ceramide would be:

Ceramide $\rightleftharpoons$ Fatty acid + sphingosine

This reaction is reversible and does not require CoA, but is probably more

compound is not only the precursor of a whole family of ceramide oligosaccharides (Table 5.1) but also stands at the branch point of two alternative pathways of ganglioside biosynthesis (Fig. 5.4).

The 'active' form of sulphate in sulphatide biosynthesis is the complex nucleotide: phosphoadenosine phosphosulphate.

The sulphatides are important constituents of the myelin sheath of nervous tissue and the incorporation of radioactive sulphate into the sulphatide molecule has been used to study the process of myelination. The 'donor' of the sulphate group is the complex nucleotide, phosphoadenosine phosphosulphate, usually abbreviated in scientific papers to PAPS.

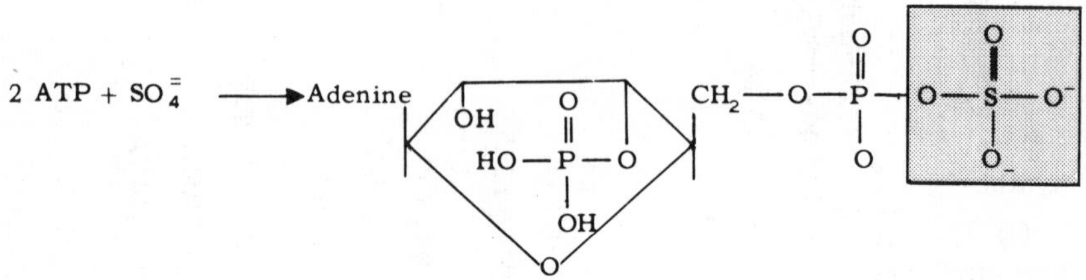

$2 \text{ ATP} + \text{SO}_4^{=} \longrightarrow$

important in the hydrolysis of ceramide than in its biosynthesis.

The more complex glycolipids, ceramide oligosaccharides, are probably formed by

The 'acceptor' of the sulphate group is a galactocerebroside and the reaction is catalysed by a soluble extract from rat brain (galactocerebroside sulphokinase):

Galactosyl-ceramide + PAPS $\longrightarrow$ Sulphatide + PAP

stepwise additions of monosaccharides by transfer from their 'active' UDP derivatives. The key substance here is *lactosyl ceramide*, formed by transfer of galactose from UDP-galactose to glucosyl-ceramide (*glucose cerebroside*). This

Although there is good evidence that the above reaction does occur, there are nevertheless indications that in brain, sulphatides may also be the precursors of monoglycosylceramides by hydrolysis of the sulphate group.

The function of glycosylceramides and sulphatides is still by no means clear, but there is a striking correlation between the sulphatide composition of some tissues and the activity of Na^+ or K^+-dependent *adenosine triphosphatase* — an important enzyme in regulating the transport of ions through membranes. The importance of lipids for the activities of membrane-bound enzymes will be discussed in chapter 6.

Gangliosides are built up by transfer of sugars to a ceramide derivative but the sequence of events is still not firmly established.

That ganglioside biosynthesis is still in its infancy can be judged from the fact that most of the communications on this subject in the scientific literature are in the form of 'preliminary notes' or 'short communications'. Some of the steps are still inferred by analogy rather than being based on firm fact. However, the following general principles of ganglioside biosynthesis seem to be clear:

(*i*) The glycosyl ceramide backbone or *part* of the backbone is formed by stepwise transfer of sugars from their UDP derivatives to ceramide.

(*ii*) Sialic acid residues are transferred from the 'active' species, cytidine monophosphate N-acetyl neuraminic acid (CMP-NANA) either before or after the backbone is completed (see Fig. 5.4).

(*iii*) The glycosyl chain is completed by further addition of monosaccharide or sialyl residues.

This picture of ganglioside biosynthesis has been largely built up in the laboratories of S. Roseman and R.O. Brady and their colleagues in the U.S.A. The way in which they worked out the sequence of sugar additions was by testing the specificity of a specific transferase for a specific 'acceptor'. Each enzyme requires a specific UDP-sugar as glycosyl donor and the *end product of the last step* has much the greatest activity as acceptor. The tentative pathways for mono, di and trisialoganglioside biosynthesis are summarized in Fig. 5.4.

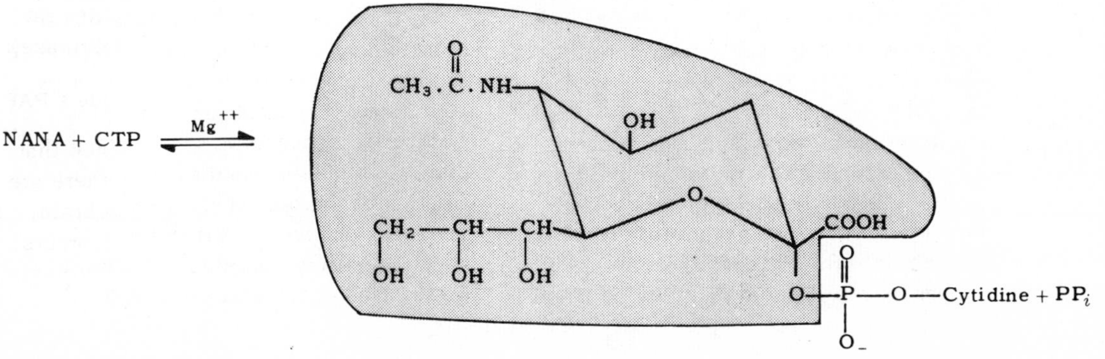

As with many other complex lipids, much less is known about the function of gangliosides than about their chemistry. The fact that in brain they are localized in just those elements which are involved in the transmission of nerve impulses suggests that they may be involved in this process, and several pieces of experimental evidence support this view. In addition, when injected into blood, gangliosides may provoke the formation of antibodies against themselves. The serological properties reside in the carbohydrate moiety and the sialic acids seem to have an essential role. The correlation between the structure of gangliosides and these suggested functions is one of the present tasks of glycolipid biochemistry.

A combination of specific enzymes can account for the complete breakdown of glycosphingolipid molecules.

Glycosphingolipids are widely distributed in animal and plant tissues and therefore constitute significant components of the human diet. Many tissues, therefore, especially the small intestinal mucosal cells, contain enzymes for breaking down glycolipids into their component parts. Usually, each of these enzymes is specific for a particular chemical bond and in combination they can account for the complete breakdown of the glycolipid molecule. Thus, for example, an enzyme has been discovered which cleaves the terminal galactose of cer-glc-gal-gal but not the galactose of cer-glc-gal. Another enzyme (from spleen) cleaves the glucose from cer-glc but is inactive on cer-gal, whereas an intestinal enzyme cleaves both glucosyl and galactosyl ceramides. But brain contains an enzyme which cleaves ceramide (but *not* cerebroside) to yield fatty acid and sphingosine. We have already mentioned that this reaction is reversible (p. 158) and that in some circumstances it could possibly be used

for synthesis. Other enzymes cleave the sulphate group from sulphatides and N-acetyl neuraminic acid from gangliosides (*neuraminidase*). Sometimes, because of a genetic defect, one or more of these enzymes is lacking and the result is a general accumulation in the tissues of that lipid which would normally be the substrate for the enzyme. This leads to a disease called a *lipidosis,* the subject of the next section.

LIPIDOSES are diseases resulting from a defect in glycosphingolipid metabolism— usually the lack of a key degradative enzyme.

The *lipidoses* proper are normally autosomal recessive inborn errors of metabolism, though subject to modification by changes in diet. Classification of these diseases is still based largely on clinical and morphological data because of lack of understanding of their detailed biochemistry. The diseases are rare and frequently fatal, an indication of the importance of maintenance of lipid structure levels, and site specificity, to the correct functioning of the organism.

When the levels of any lipid are increased to such an extent that they are deposited in tissues normally free of them, the cause is either a lack of one or more degradative enzymes due to gene mutation, or the failure of a regulatory gene (not necessarily only in lipid metabolic sequences). The rarity of the diseases hinders progress in indentification of the biochemical steps involved, and we are therefore in this section limiting ourselves to a brief description of the disease and its associated lipid abnormality.

It might be noted that we have less detailed knowledge of human lipid metabolism than that of any other biological group. This hinders clinical research in many fields and can be recommended to students seeking a difficult research area in which to make their mark.

The first of the lipidoses which we will describe involves the phosphosphingolipid, sphingomyelin (see chapter 4). This is not a glycolipid and by that criterion, a description of *sphingomyelinosis* would be out of place here. However, the disease is so much akin to the other glycosphingolipidoses that it is best dealt with in this section.

(1) Sphingomyelinosis (Niemann—Pick Disease)

Clinical symptoms. This is a rare inherited condition characterized by deposition of sphingomyelin in almost every organ and tissue, but particularly in the endothelial, mesenchymal, and parenchymal cells. The chemical signs are usually visible in the first year of life and the disease is frequently fatal before the third year. No treatment is known.

Enzymic defects. The lesion is due to a deficiency of the enzyme catalysing sphingomyelin cleavage.

$$\text{Sphingomyelin} + H_2O \xrightarrow{\text{Sphingomyelinase}} \text{ceramide} + \text{phosphoryl choline}$$

The fatty acid composition of the sphingolipids in diseased tissue is similar to the foetal pattern (i.e. more stearic and less longer chain acids), as is frequently found in other lipidoses that involve the central nervous system. That the metabolic error involved sphingomyelin *cleavage* rather than *synthesis* was confirmed by D.S. Frederickson, who showed that the activity of the cleavage enzyme in livers and kidneys of patients was about 7% of that of normal subjects.

(2) Gangliosidoses

The first systematic study of the chemistry of the gangliosides was done by Klenk in Germany and arose from study of Tay—Sachs disease, a familial condition involving accumulation of a then unknown glycolipid.

(a) Tay—Sachs disease

Clinical symptoms. This is the commonest *gangliosidosis* with the following characteristics: regression of mental and somatic function; increasing impairment of vision; lipid accumulation in ganglion cells, and demyelination.

Enzymic defect. Tay—Sachs ganglioside ($G_{GNTrII}1$ or G_{M2}) — the ganglioside lacking the normal terminal galactose — accumulates in the brain, but the specific metabolic defect has been difficult to locate. This is due to the difficulties in synthesizing labelled Tay—Sachs ganglioside. The missing enzyme is probably the one which hydrolyses the terminal N-acetylgalactosamine of the Tay-Sachs ganglioside.

(b) Neurovisceral gangliosidosis

This is differentiated from Tay-Sachs disease by storage of gangliosides (particularly G_{GNT^1} or G_{MI} in visceral organs as well as in brain. The symptoms, however, are similar to those of Tay—Sachs disease.

(3) Lipidoses involving glycosyl ceramides

(a) Gaucher's disease

Clinical symptoms. This is probably the commonest lipid storage disease. The main clinical characteristics are hepatomegaly, splenomegaly, skin pigmentation, and chronic progression. There are two forms. In the infantile form, but not in the adult form, there is involvement of the central nervous system. The disease is also characterized by the appearance of large lipid-laden cells in spleen, liver, and bone marrow.

Enzymic defect. The lipid which accumulates in these cells is glucosyl ceramide and the metabolic abnormality is a deficiency of the enzyme, glucocerebrosidase : —

$$\text{Glucosyl ceramide} + H_2O \xrightarrow{\text{glucocerebrosidase}} \text{glucose} + \text{ceramide}$$

The main source of the glucosyl ceramide is lactosyl ceramide, the major lipid of leukocytes.

(b) Fabry's disease

Clinical symptoms. The onset of this disease is demonstrated by a localized skin rash around the age of ten and then by pains in the extremities and pyrexia. There is then progressive renal failure and death ensues between the ages of 30 and 60. The full complement of symptoms is seen only in males, women having an incomplete form.

Enzymic defect. Trihexosyl ceramide accumulates, particularly in the kidney, due to a deficiency of the enzyme which splits the terminal galactose unit from trihexosyl ceramide.

$$\text{Cer-glc-gal-gal} \xrightarrow[\text{trihexosidase}]{\text{Ceramide}} \text{lactosyl ceramide} + \text{galactose}$$

(c) Metachromatic leucodystrophy

Clinical symptoms. This is an inherited disorder of sulphatide metabolism involving the myelin of the central nervous system. The disease can manifest itself at any age and is characterized by impairment of motor function, ataxia, coarse tremor and progressive speech impairment. In children there is arrest of development and loss of acquired skills. The condition is inexorably progressive and results in death after extensive demyelination of the central nervous system has occurred.

Enzymic defect. The lipids which accumulate are the sulphatides and the biochemical abnormality is a deficiency in the sulphatase which converts sulphatides into cerebrosides :

$$\text{Sulphatide} + H_2O \xrightarrow{\text{Sulphatase}} \text{galactosyl ceramide} + \text{sulphuric acid.}$$

Table 5.3. Summary of Biochemical Defects in the Sphingolipidoses

Name of disease	Lipid which accumulates	Missing enzyme
Sphingomyelinosis (Niemann—Pick)	Sphingomyelin	Sphingomyelinase
Gangliosidoses		
(a) Tay—Sachs	G_{GNTrII}[1]	Specific N-acetyl galactosidase
(b) Neurovisceral Gangliosidosis	G_{GNT}1	Specific β-galactosidase
Cerebrosidoses		
(a) Gaucher	Cer-Glc	glucocerebrosidase
(b) Fabry	Cer-Glc-Gal-Gal	Specific galactosidase ceramide trihexosidase
(c) Metachromatic leucodystrophy	Sulphatide	Sulphatase

At the present time, there is no specific therapy for any of the lipid storage diseases, although now that the metabolic defects have in many cases been identified, several exciting possibilities are in sight. There is a good chance that the enzyme deficiency responsible for many of these diseases can be detected *in utero*. Once the deficiency has been established, *enzyme replacement* is not beyond the bounds of possibility or alternatively, administration of the appropriate messenger RNA or specific DNA molecules. This kind of enzymological diagnosis followed up by *'genetic engineering'* will now put a new and heavy responsibility on the shoulders of the biochemist.

GLYCOSYL GLYCERIDES

Glycolipids based on glycerol have a structure analogous to phospholipids with the sugar attached glycosidically to the 3-(sn)-position of glycerol and fatty acids at the 1- and 2-(sn) positions.

In contrast to the glycosyl ceramides which are important mainly in animals, the glycosyl glycerides are major constituents of plants and microorganisms, although small quantities have been detected in animals, mainly in brain. The most important of these are classified in Table 5.4. In all these compounds the glycerol has the same configuration as in the phospholipids. According to the old nomenclature these would have been called: 1-glycosyl-2,3-diacyl-D-glycerols but in the more satisfactory stereochemical numbering system they become: 1,2-diacyl,3-glycosyl-(sn)-glycerols. A large number of sugars may

Table 5.4. Glycolipids based on glycerol (glycosyl glycerides)

Common name	Structure and chemical name
Monogalactosyl diglyceride (MGDG)	1,2-diacyl-[β-D-galactopyranosyl(1'→3)]-*sn*-glycerol
Digalactosyl diglyceride (DGDG)	1,2-diacyl[α-D-galactopyranosyl-(1'→6')-β-D-galactopyranosyl(1'→3)]-*sn*-glycerol
Plant sulpholipid (sulphoquinovosyl-diglyceride)	D-quinovose is 6-deoxy-D-glucose Note the carbon-sulphur bond. 1,2-diacyl-[6-sulpho-α-D-quinovopyranosyl-(1'→3)]-*sn*-glycerol.
Mannosyl diglycerides (and others)	1,2-diacyl[α-D-mannosyl(1'→3')α-D-mannosyl (1'→3)]-*sn*-glycerol

164

Especially abundant in plant leaves and algae; mainly in chloroplast. Contains a high proportion of polyunsaturated fatty acids. *Chlorella vulgaris* MGDG has mainly 18:1, 18:2 when dark grown but 20% 18:3 when grown in the light. *Euglena gracilis* MGDG has 16:4. Spinach chloroplast MGDG has 25% 16:3, 72% 18:3. Also found in the central nervous systems of several animals in small quantity.

Usually found together with MGDG in chloroplasts of higher plants and algae. Not quite so abundant as MGDG. Also has high proportion of polyunsaturated fatty acids, especially 18:3. In both lipids the glycerol has the same configuration as in the phospholipids (see chapter 4). The naming is confusing; many authors still use the 'D-glycerol system and you will see the galactolipids written: β-*D*-galactopyranosyl (1→1′)2′,3′-diacyl-*D*-glycerol (MGDG) and α-*D*-galactopyranosyl (1→6)β-*D*-galactopyranosyl (1→1′)2′,3′-diacyl-*D*-glycerol (DGDG).

Usually referred to as a 'sulpholipid' as distinct from a 'sulphatide' which is reserved for cerebroside sulphates. Found in leaves and algae. Contains more saturated fatty acids (mainly palmitic) than the galactolipids, e.g. spinach leaf sulpholipid has:
27% 16:0, 39% 18:2, 28% 18:3.
Glycerol has the same configuration as in the galactolipids and phospholipids. The alternative name would be 6-sulpho-α-*D*-quinovopyranosyl(1→1′)2′,3′-diacyl-*D*-glycerol.

Found in bacteria especially the micrococci. The di-mannosyl derivative is most abundant but a monomannosyl derivative also occurs. *Pneumococci* have galactosyl-glucosyl-diglycerides and a variety of other structures. *Streptococci* and *Mycoplasma* have mono- and diglucosyl derivatives.

be attached glycosidically to the 3-(sn)-position of glycerol. In higher plants and algae the most abundant sugar is galactose, whereas mannose and glucose are more often found in the bacteria. 6-Sulphoquinovose is found as a constituent of the sulpholipids of all photosynthetic plants, algae and bacteria so far examined. The fatty acids of plant glycolipids are highly unsaturated while those of bacteria contain a large proportion of branched chain fatty acids (see Table 5.4).

Galactosyl diglycerides are concentrated in the chloroplasts of higher plants and algae but their concentration is much reduced when the cells become etiolated.

Galactosyl diglycerides and the sulpho-quinovosyl diglyceride are frequently referred to as 'chloroplast lipids'. They make up the bulk of the glycerolipids of chloroplasts; phosphatidyl glycerol is the only phospholipid present in significant quantity. Little careful analytical work has been done on plant subcellular fractions, however, so that their presence in other particles such as mitochondria cannot be excluded. When green plant tissue is deprived of light, the chlorophyll content diminishes, chlorplast structure breaks down and the tissue becomes white or 'etiolated'. A particularly useful organism for studying such changes is the protist, *Euglena gracilis*, which behaves completely like a plant when grown in the light on purely inorganic salts, but in the dark becomes etiolated and will live like an animal as long as suitable organic compounds are supplied

to it. The process can be reversed again by exposing the cells to light and exchanging the organic for an inorganic medium. The gradual appearance of chlorophyll and organized chloroplast structure can be observed, and parallel with it, the synthesis of galactolipids and sulpholipid. Since polyunsaturated fatty acids such as α-linolenic acid ($^{9,12,15}C_{18:3}$) are esterified mainly in these lipids and are also formed only in the light, their synthesis also parallels that of the glycolipids and chlorophyll. Most organisms in which photosynthesis is accompanied by evolution of oxygen contain large quantities of α-linolenic acid whereas the more primitive organisms which do not evolve oxygen during photosynthesis, do not have this acid. This fact, together with the close correlation between galactolipids, linolenic acid and 'photosynthetic competence' led to the idea that galactolipids and linolenic acid in particular played some part chemically in the oxygen-evolving step of photosynthesis. Photosynthetic organisms do exist, however, which can evolve oxygen but do not possess linolenic acid, so that the subject is still controversial.

Galactosyl diglycerides are formed by transfer of the sugar from UDP-galactose to a diglyceride

The biosynthesis of galactosyl diglycerides has been difficult to demonstrate *in vitro* and studies in this area are relatively new. UDP-galactose* had long been assumed to be the sugar 'donor' but it was some time before isolated

* Uridine diphosphate galactose.

chloroplast preparations could be made which would catalyse the incorporation of ^{14}C-galactose from its sugar nucleotide into galactolipids. The nature of the 'acceptor' was not demonstrated until A. Ongun and J.B. Mudd in California showed that an acetone powder of spinach chloroplasts would catalyse the incorporation of galactose from UDP-galactose into monogalactosyl diglyceride only if the lipids which had been extracted by the acetone were added to the incubation mixture. The 'acceptor' for monogalactosyl diglyceride synthesis proved to be a diglyceride; for digalactosyl diglyceride synthesis, only naturally occurring monogalactolipid could act as an acceptor. The enzymes which catalyse each of these galactose transfers seem to be quite distinct:

Table 5.5. Fatty acids of the Galactolipids of Light-grown *Euglena gracilis*.

	MGDG	DGDG
	(% of total fatty acids in each lipid class)	
16:0	6	17
16:4	32	7
18:1	9	19
18:2	6	12
18:3	41	26
others	6	19

cation of the lysogalactolipids. Recent work, too, especially with phospholipids, has demonstrated that some enzymes are capable of 'recognizing' certain molecular species, so that either of these alternatives is quite possible.

The pathways by which mannosyl diglycerides are formed in the bacterium,

1,2-Diglyceride + UDP-galactose ⟶ Monogalactosyl diglyceride

Monogalactosyl diglyceride + UDP-galactose ⟶ Digalactosyl diglyceride.

Although chloroplasts are the main source of galactosyl diglycerides, brain also contains an enzyme which catalyses the formation of monogalactolipid from diglyceride and UDP-galactose.

A puzzle which remains to be solved is how and why the differences in fatty acid composition between monogalactosyl and digalactosyl diglyceride arise. These differences are quite marked in *Euglena gracilis* (see Table 5.5). Are certain molecular species of monogalactosyl diglycerides 'selected' by the enzyme which adds the second galactose, or is the fatty acid composition modified after the formation of digalactosyl diglyceride? We *do* know that enzymes exist in spinach leaves which remove fatty acids and catalyse the re-esterifi-

Micrococus lysodeikticus are very similar except that the guanidine nucleotide, GDP-mannose, is the sugar donor.

The biosynthetic pathway leading to the sulpholipid is obscure. It seems to be only speculation that UDP-sulphoquinovose donates the sulpho-sugar to diglyceride in an analogous way to the biosynthesis of other glycosyl glycerides. More attention has been given to the biosynthesis of the sugar moiety, but even then, the origin of the *carbonsulphur* bond is far from certain.

Enzymes exist for the complete degradation of glycosyl glycerides into their component parts.

Enzymes have been found in spinach leaves and runner beans which hydrolyse the fatty acids from mono- and digalactosyl diglycerides to form galactosyl and digalactosyl glycerol. A combination of α- and β-galactosidases can then completely degrade these compounds to galactose and glycerol. These degradative enzymes are so active that galactolipids often cannot be detected in leaves which have been homogenized in water; the enzymes act immediately the cells are broken. *Lyso* compounds are probably formed as intermediates and these may be reacylated to the diacyl lipid by transfer of fatty acids from acyl-CoA, thereby accounting for the fatty acid 'turnover' of glycolipids in leaf tissue.

The function of glycolipids is still obscure.

A favourite pastime of lipid biochemists is to speculate about the function of lipids in cells. Apart from the possibility of their participation in the mechanism of photosynthesis, glycolipids have been suggested as 'carriers' of sugars across cell membranes. There is little evidence for these views. The authors' opinion is that the physical properties resulting from the interaction of the polar and non-polar groupings of a particular lipid mixture with the proteins in a membrane are especially suitable for the particular membrane in which they are situated.

OTHER GLYCOLIPIDS

Other glycolipids are:
In bacteria,
(i) Esters of carbohydrates
(ii) Glycosides of fatty alcohols
and hydroxyacids
In Algae: Chlorosulpholipids
In higher plants: Esterified
sterol glycosides.

There remains a whole series of glycolipids which do not fit into either of the classes we have so far discussed. These are mainly complex bacterial lipids whose structure is in many cases uncertain. We can only deal briefly with some of these and the reader is referred to the writings of E. Lederer, the French chemist who is one of the foremost authorities on the subject (see bibliography at the end of the chapter).

Esters of carbohydrates

These constitute a whole family of complex glycolipids mainly confined to the *Mycobacteria.*

(a) Cord factor

This is an ester of the disaccharide, trehalose, with two molecules of the complex fatty acid, *mycolic acid.* This acid is not a single chemical species but the term embraces a whole series of fatty acids containing 60–90 carbon atoms. They are hydroxy acids with differing degrees of unsaturation and

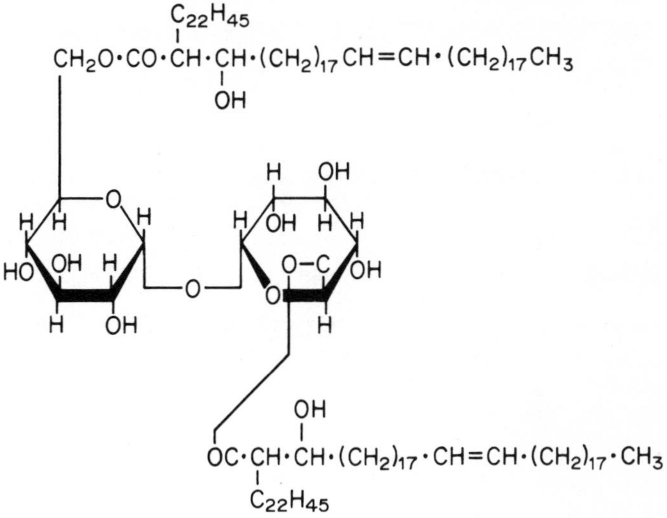

Fig. 5.5. Cord factor.

chain branching. An example is the C_{60} mycolic acid of *Mycobacterium smegmatis*, shown in Fig. 5.5. Cord factor is so called because it is found in the waxy capsular material of virulent strains of tubercle bacilli and related bacteria. This causes the bacteria to string together in a long chain or cord. The compound is highly toxic and is somehow intimately associated with the virulence of the organism. Little or nothing is known about its biosynthesis.

(b) Waxes A, B, C and D.

These are esters of very long chain fatty acids with high molecular weight polysaccharides and peptides, and so really belong to chapter 6 where they are described briefly.

(c) Esters of inositol.

Propionibacteria contain diacyl myoinositol mannosides in which the mannose is glycosidically linked to position 2 of myoinositol and the acyl groups esterified at positions 1 and 6 of the inositol ring (see Table 4.1 for numbering of inositol). Other bacteria contain esters of glucose and certain other sugars.

Glycosides

(a) Mycosides.

Mycosides are again characteristic of the *Mycobacteria*. The basic structure is a long chain, highly branched, hydroxylated hydrocarbon terminated by a phenol group, with the sugar linked glycosidically to the phenolic hydroxyl. Acyl chains are

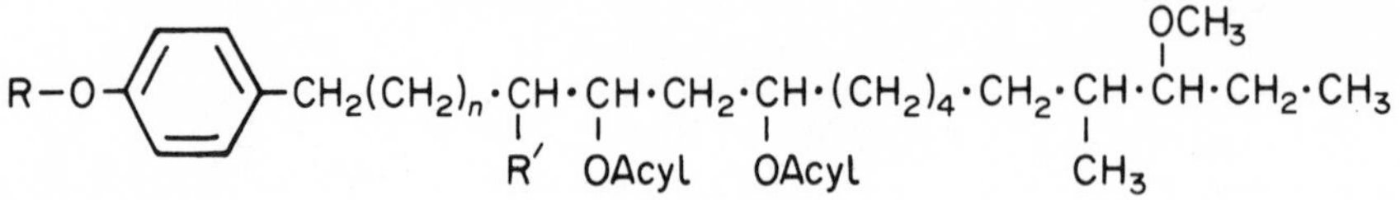

R–O–⟨phenyl⟩–$CH_2(CH_2)_n \cdot CH \cdot CH \cdot CH_2 \cdot CH \cdot (CH_2)_4 \cdot CH_2 \cdot CH \cdot CH \cdot CH_2 \cdot CH_3$

with substituents: R', OAcyl, OAcyl below the chain; OCH_3 above and CH_3 below at the right end.

$n = 13–17$

R = Trisaccharide containing: 2–O–methyl fucose
2–O–methyl rhamnose
2,4–di–O–methyl rhamnose

$R' = CH_3$ or H

Acyl = normal saturated fatty acids, 12:0, 14:0, 16:0,
or 18:0, tuberculostearic acid, mycocerosic, 22:0, 29:0, 30:0, 32:0,

Fig. 5.6. Structure of mycosides.

esterified to hydroxy groups of the hydrocarbon chain (Fig. 5.6).

(b) Esterified sterol glycosides

These occur in plants and have been described in the section on sterol esters (chapter 3, page 113).

Chlorosulpholipids

Naturally occurring covalently bound organic chlorine is extremely rare in biological tissue – although *artificial* organic chlorine compounds such as insecticides are quite commonly found nowadays. Until recently, only fungi were known to contain chloro compounds, such as the chloroketoacid *caldariomycin* $(Cl \cdot CH_2 CO \cdot CH_2 \cdot CH_2 \cdot COOH)$. In 1969. two research groups reported on the presence of a chlorosulpholipid in the lipids of the green alga *Ochromonas danica*. The compounds were detected when the organism was grown in the presence of $Na_2^{35}SO_4$ and $Na^{36}Cl$ so that the radioisotopes became incorporated into the lipids. There is a family of these lipids with two sulphate ester groups and from one to six chlorine atoms. An example of dichlorolipid is shown below.

The analysis of glycolipid mixtures:

As in the case of the other lipids we have discussed (Neutral Lipids, chapter 3; Phospholipids, chapter 4) useful structural and biosynthetic work can only be done if adequate techniques have been developed for the isolation and purification of the different types of glycolipids. These do

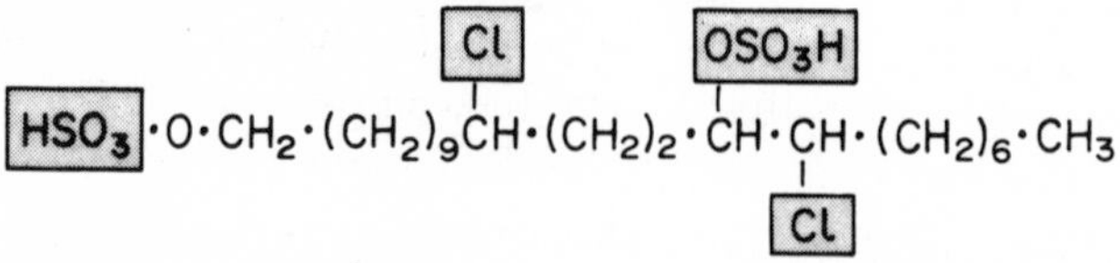

$\boxed{HSO_3} \cdot O \cdot CH_2 \cdot (CH_2)_9 CH \cdot (CH_2)_2 \cdot CH \cdot CH \cdot (CH_2)_6 \cdot CH_3$

with $\boxed{Cl}$ above, $\boxed{OSO_3H}$ above, and $\boxed{Cl}$ below.

Fig. 5.7. **11,15-dichloro-docosan-1, 14-disulphate.**

not differ in principle from those already described for other lipids; chloroform—methanol extraction and washing procedures are fairly general. When large quantities of lipid are required, as for instance in structural analysis, column chromatography on silicic acid has been the most widely used method. However, DEAE-cellulose chromatography has been very useful for separating plant glycolipids. It does no harm to emphasize yet again that complete purification can only be achieved by a combination of methods. A final 'clean-up' procedure would almost certainly be thin layer chromatography. This has also become the most widely used general technique especially as structural studies have largely been completed and biosynthetic work with radioactive tracers, which needs far less material, have taken their place. Successive TLC runs in several different solvents or two dimensional TLC in which the plate is turned through 90° after development in the first solvent can usually provide sufficient resolution. In the analysis of closely related glyco-lipids differing only in the type of sugar in the molecule, TLC on silicic acid which has been impregnated with boric acid has been very productive. Such compounds as glucosyl and galactosyl

ceramides may be resolved by this technique.

Individual components of the glycolipid molecule derived by hydrolysis are analysed by standard chemical methods or increasingly by physical methods such as mass spectrometry.

Acid or alkaline hydrolysis of the purified products yields components which can be analysed by the standard methods of fatty acid or sugar chemistry. GLC is widely used nowadays for sugars in the form of their trimethyl silyl (TMS) ethers and a great deal of analytical work on sphingo-sine bases has been done by GLC. Another example of the increasing use of physical methods for the analysis and structural identification of unknown compounds is mass spectrometry. Many of the complex glycolipids of the Mycobacteria have been analysed by Lederer with the aid of mass spectrometry and this technique was also used for the determination of the structure of the chlorosulpholipids. Modern tendencies have been to combine GLC separation with mass spectroscopic analysis and to feed the data directly into a computer. This promises to become a general procedure especially for routine analyses.

SUMMARY

The majority of naturally occurring glyco-lipids contain the sugar linked by a glycosidic bond to either the 1-hydroxyl group of an N-acyl sphingosine (ceramide) or to the 3-(*sn*)-position of a diglyceride.

Glycosyl ceramides are widely distri-buted in both animals and plants but are most abundant in animal brains. If they contain one sugar, they are called *cerebrosides;* this sugar is usually galactose but occasionally glucose. *Cerebroside sulphates* (sulphatides) contain a sulphate ester at the 3-position of the sugar. In *ceramide oligosaccharides* several sugars are linked glycosidically to each other, while in *gangliosides*, the oligosaccharide chain also contains 1-3 molecules of N-acetyl neuraminic acid (sialic acid). The fatty acid is in all cases bound to the amino group of sphingosine by an amide bond.

There are alternative pathways for the biosynthesis of each of these compounds, the relative importance of which has not been resolved. *Cerebrosides* may be formed either by acylation of glycosyl sphingosine with acyl-CoA or by transfer of a sugar to acyl sphingosine. The active form or 'donor' of the sugar is always the UDP-sugar whereas the donor of sialyl residues is CMP-N-acetyl neuraminic acid (CMP-NANA). Sulphatides are formed by transfer of sulphate from phosphoadenosine phosphosulphate to a cerebroside. There are enzymes specific for the hydrolysis of each type of bond in glycosyl ceramides so that they may be biologically degraded into their component enzymes, lipid may accumulate giving rise to a *lipidosis*. *Glycosyl glycerides* are also widely distributed but are most abundant in chloroplasts of algae and higher plants chloroplasts of algae and higher plants where the sugar is galactose. Both mono- and digalactosyl diglycerides occur. In bacteria a wider variety of sugars is found, including glucose, galactose and mannose. These compounds are formed by transfer of sugars from the sugar nucleotide (usually UDP-sugar) to a 1,2-diglyceride 'acceptor'. In chloroplasts, a sulpho- quinovosyl diglyceride is found in which the sulphur is present in a *sulphonic acid* group not a sulphate ester.

Bacteria also contain a wide range of complex glycolipids which are either esters of carbohydrates with long chain, highly branched, hydroxy fatty acids or glycosides of fatty alcohols or fatty acids.

Recently, a whole new range of sulpho- lipids containing organic chlorine has been discovered in certain green algae.

BIBLIOGRAPHY

General

1. CARTER H.E., JOHNSON P. and WEBER E.J. (1965). Glycolipids. *Ann. Rev. Biochem.* 34, 109. A comprehensive review of most glycolipid types but not completely up to date now.

Plant Glycolipids

2. NICHOLS B.W. and JAMES A.T. (1968). Acyl lipids and Fatty Acids of Photosynthetic Tissue. *Prog. Phytochem.* 1, 1. Edited by L. Reinhold and Y. Liwschitz, Interscience, London.

3. ROSENBURG A. (1967). Galactosyl Di- glycerides: their possible function in *Euglena* chloroplasts. *Science N.Y.* 157, 1191.

4. BENSON A.A. (1963). The plant sulpho- lipid. *Advances in Lipid Research* 1, 387. Very little extra work on sulpholipid bio- synthesis has appeared since this article.

Glycosphingolipids and lipidoses

See Ref. 1 above.

5. BRADY R.O. (1969). Genetics and the sphingolipidoses. *Medical Clinics of North America* 53, 827.

6. WIEGANDT H. (1968). The structure and function of gangliosides *Angew. Chem.* (International Edition) 7, 87.

Bacterial Glycolipids

7. LENNARZ W.J., (1966). Lipid metabolism in the bacteria. *Advances in Lipid Research* 4, 175.

8. LEDERER E. (1967). Glycolipids of mycobacteria and related microorganisms. *Chemistry and Physics of Lipids* 1, 294.

9. ASSELINEAU J. (1962). Les Lipides Bacteriens. Hermann, Paris. The 'Bible' of bacterial lipid biochemistry.

6 Lipids as components of macromolecules

In earlier chapters we have dealt with the metabolism of fatty acids and lipids considered as individual molecules. In fact, as we have pointed out several times, they are rarely present in cells as 'free' lipid molecules and we must now consider them as they really are — part of a macromolecular array, either in combination with protein or with carbohydrate. This fact is readily apparent to the research worker when he first tries to extract lipids from a tissue with what is normally a quite adequate lipid solvent — ether or chloroform. Only when a more polar solvent such as an alcohol is added can the full complement of lipids be removed and we can infer that some physical-chemical association is first being disrupted before the lipids can move into the completely non-polar phase. In this chapter we will first examine the chemical and physical properties of lipid and protein molecules and try to decide what kind of association would be expected. We will then describe the various types of lipid-containing macromolecules in living cells ranging from the relatively small 'soluble' lipoprotein molecules to the three dimensional aggregations of lipid and protein or lipid and carbohydrate which constitute the cell membrane or cell wall. These membranes are not merely 'structural' but contain many enzymic proteins for whose activity a completely integrated macromolecular structure is often necessary. We will therefore investigate the part which the lipids might play in preserving enzymic activity in this three dimensional *milieu*.

1. PHYSICAL AND CHEMICAL PROPERTIES OF LIPIDS

*An examination of the chemical
groups of lipids to assess the
possibilities of bonding with
other molecules. The importance
of non-polar, polar and charged
groups.*

The C–C and C–H bonds which constitute
the hydrocarbon chains typical of fatty
acids have a rather symmetrical electron
distribution around them. There are no
areas in which an electrical charge can
be localised and we refer to these compounds as *non-polar* or *apolar*. The only
forces which can hold such molecules
together are those which cause a weak
attraction of all bodies for each other,
namely *Van der Waals* or *London Dispersion* forces; they are weak and operate
only over short distances. We often refer
to the forces which hold non-polar
molecules together in this way as *hydrophobic bonds*. Good examples of lipids
having predominantly non-polar groups
are long chain saturated fatty acids and
esters, and the fused ring system of
steroids such as cholesterol. When two
atoms are present which have significantly
different electronegativities, or electron
attracting power, such as carbon and
oxygen, there will be an unsymmetrical
electron distribution in the chemical bond
as electrons will tend to accumulate
around the more electronegative atom.
Groups of this sort are said to be *polar*.
The best example of a ubiquitous polar
biological molecule is water. The partial
negative character of the oxygen may
attract the slightly positive hydrogen of
another water molecule to form a rather
weak $O - H ... O$ bridge. Such bridges, in
which a hydrogen atom comes between two
electronegative atoms such as oxygen-
oxygen, nitrogen-nitrogen or oxygen-
nitrogen, are called *hydrogen bonds*. In
the case of water, the oxygen atom is
capable of hydrogen bonding with two
hydrogens, one from each of two other
water molecules, and in this way, a tetrahedral structure is built up, over fairly
short distances in liquid water, but
infinitely in the case of ice. In lipid
molecules the oxygen of hydroxyl,
carbonyl or ether groups, and the nitrogen
of amino groups, for example, can take
part in hydrogen bond formation. In certain
cases, a bond may not be polar but have
the 'potential for polarity' or be
polarizable. A case in point is the double
bond of an unsaturated fatty acid in which
the electrons may be polarized by a
charged species. Further up the scale
of polarity are the permanent dipoles,
such as the ester bonds in glycerides and
phospholipids, ether bonds, and the
hydroxy groups of mono- and diglycerides,
lysophospholipids and inositides. In the
extreme case we have a complete
separation of charge. Examples of
negatively charged groups are the phosphate and sulphate of phospholipids and
sulpholipids and the carboxyl of phosphatidyl serines. Positive charges are
found in the quaternary ammonium of
choline phospholipids and the primary
amine groups in phosphatidyl serines or
ethanolamines.

The presence of water molecules is important in determining the way in which lipid molecules interact.

Having examined some of the chemical groupings in lipid molecules, what can we say about the way in which lipids will react with each other and with their environment, bearing in mind that in biological systems this environment will usually, but not always, be aqueous? Completely non-polar molecules have no groups which can associate with the water molecules in the medium. They tend, therefore, to associate with each other by *Van der Waals* forces into spherical droplets whose surface area is minimal. At higher concentrations, the tendency is for these droplets to coalesce to larger ones so that the interfacial area is as small as possible. Triglycerides behave in this way, because although the ester groups contain permanent dipoles, the hydrocarbon chains are so preponderant as to render the molecules almost completely non-polar. If polar lipid (or protein) molecules are introduced into the triglyceride-water system they will tend to arrange themselves with their lipid groups buried in the triglyceride droplets but their polar groups at the surface. The effect of these so-called *surface active agents* is to prevent the droplets from coalescing and to create a more stable dispersion of the fat, usually termed an *emulsion*.

When polar lipids alone are dispersed in water, they will tend to orientate themselves so that their charged or strongly polar groups associate with the water molecules, while their non-polar 'tails' associate with each other. Single molecules exist only to a very

limited extent. The molecules prefer to form aggregates, usually called *micelles* which are in equilibrium with single molecules. How far this equilibrium favours the existence of single molecules depends on the relative proportions of polar and non-polar groups in the molecule. However, as the concentration is raised, a characteristic concentration is reached known as the *critical micellar concentration* above which the addition of more of the *amphipathic* substance (amphi = of both kinds; pathos = feeling) increases the number and size of the micelles but causes a relatively small increase in the concentration of single molecules. The term *micelle* is generally used for the spherical or cylindrical aggregates (see Fig. 6.1) which are formed when the lipid content is fairly low relative to the aqueous phase. When the *water* content is low, on the other hand, the lipids can exist in a number of *liquid crystalline phases*. These have been studied with the aid of X-ray diffraction techniques by Dr. V. Luzatti's research group in France and *a few* are illustrated in Fig. 6.1. The predominant phase depends on such factors as the water content, temperature and the type of lipid involved.

For example, a zwitterionic phospholipid such as phosphatidyl choline more readily forms sheet-like arrays or *lamellae* [Figs. 6.1(a), 6–4(c)] because the positive and negative charges on the headgroup balance each other and there is therefore no repulsion between them. This allows the hydrocarbon chains to come close together to achieve the maximum degree of hydrophobic bonding. The cross sectional area of two fatty acid chains, which are roughly parallel and the phosphoryl choline moiety are approximately equal and so the molecule is cylindrical — another point which favours formation of lamellae. Strongly acidic phospholipids, however, such as the phosphoinositides will experience considerable repulsion between headgroups and tend to more readily form spherical micelles. These are generalizations, however, and we should point out that in some circumstances, lipids with no overall charge do form spherical micelles, while charged lipids do sometimes form lamellae. X-ray diffraction can be supplemented by light and electron microscopic studies to find out more about the different phases adopted by lipid molecules under different conditions.

Numerous physical techniques are employed to measure the forces of inter-action between lipids. One of the oldest of these is the measurement of force-area characteristics of monomolecular films. A small amount of substance is allowed to spread on the surface of water and its 'compressibility' measured by suitable apparatus. Films of long chain saturated fatty acids at an air-water interface are 'condensed' at relatively low pressures because the hydrocarbon chains can pack together as in a crystal-line substance near its melting point. The hydrocarbon chains are then almost normal to the surface. Films of highly charged substances reflect the repulsive effects of the charged groups and the films are more expanded. Similar effects are brought about by introducing double bonds into the hydrocarbon chain. In this case the 'kink' in the chain does not allow such close packing. By studying the change in compressibility of films brought about by the addition of other substances such as cholesterol, a good insight into the degree of interaction of different lipid molecules can be gained.

A more recent tendency has been to employ various spectroscopic techniques. To view a spectrum is essentially to get a picture of molecular and atomic motion. Infrared spectra can tell us about the rotation of certain groups in molecules while NMR is being used increasingly

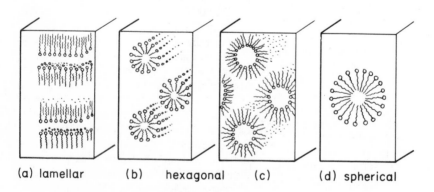

(a) lamellar   (b) hexagonal   (c)   (d) spherical

Fig. 6.1. Some possible lipid-water systems.

to measure the amount of mobility of lipid moieties containing characteristic proton groupings — for example the $-CH_2-$ protons of the hydrocarbon 'tails' or the $-CH_3$ protons of the choline 'head-groups'. These methods (and many others such as *differential thermal analysis*) tell us that over a certain range of temperature the hydrocarbon chains of a phospholipid can flex and twist and be highly mobile. The more unsaturated the chain, the lower the temperature at which this mobility is apparent. However, this mobility cannot occur if the molecules are constrained, for example by reaction with other molecules such as water or protein. Physical methods, then, can tell us something about the interactions of whole, or parts of, lipid molecules with each other or with other types of molecules in their environment.

2. PHYSICAL AND CHEMICAL PROPERTIES OF PROTEINS

Weak bonds, similar to those which hold lipids together in micelles are responsible for maintaining the characteristic conformations of single protein molecules.

While individual lipid molecules often tend to form large three-dimensional arrays as we have just described, a single protein molecule can be thought of as an array of amino acids linked covalently in a linear manner but arranged finally in a three-dimensional structure which is stabilized by inter-actions between amino acid residues in different parts of the chain. The same kinds of interactions occur as we described for lipid molecules. Non-polar groupings are exemplified by the side chains of valine, leucine or isoleucine. An example of a polarizable group might be the aromatic double bond system of phenylalanine, while permanent dipoles can be found in the hydroxyl, sulphydryl and amide groups of serine, cysteine and asparagine respectively. At the extreme ends of the 'polarity scale' are the carboxyl groups of aspartic or glutamic acids with a negative charge and the imidazole, ammonium or guanidinium groups of histidine, lysine and arginine respectively, with positive charges. Changes in environment which might cause aggregation of lipids into larger micelles or conversely cause dissociation into smaller particles or free molecules, may, in a protein, cause alterations in the interactions between one part of the chain and another, resulting in a change of position of amino acid residues or side chains relative to each other in space. The freedom of movement of individual residues is severely restricted by the existence of peptide linkages between each constituent of the chain, but the presence of the $-CO-NH-$ back-bone presents opportunities for hydrogen bonding between different parts of the chain. The interactions which hold proteins in certain characteristic con-figurations (sometimes helical) are therefore the rather weak forces of hydrogen bonds and *Van der Waals* interactions similar to those existing between lipid molecules in micellar aggregates; (a notable exception are the covalent —S—S— 'bridges' which often link different protein chains together). The predominance of weak bonds is reflected in the extreme sensitivity of protein molecules to heat and other 'denaturing' agents. These weak interactions may occur not only between different parts of the same chain, but between different peptide chains which become associated with each other. Many enzyme proteins exist as an association of *sub-units* and an example of such an enzyme in the realm of lipid metabolism — *Thiolase* — has already been described in chapter 2.

3. ASSOCIATION OF LIPIDS AND PROTEINS

We have shown that the forces which bind individual lipid molecules together in aggregates and those stabilizing large protein molecules and maintaining the secondary or tertiary structure, are of a similar kind. We know that discrete chemical complexes containing lipids and proteins, which we call lipoproteins, exist in cells. The fact that these can be isolated and characterized indicates that there is a definite interaction between the two moieties. What, then, are the predominant forces which bind lipids to proteins in living cells? It is generally accepted that covalent bonds between lipid and protein are unimportant and indeed many people consider that they can be ruled out entirely because in most cases the lipid can be extracted with organic solvents under mild conditions. Nevertheless, some examples of fatty acids covalently bonded to protein do seem to be genuine. The main possibilities then, lie between hydrophobic forces at one end of the scale and hydrogen bonds or electrostatic forces at the other. We have to admit that the answer to this question is still uncertain in most cases and we believe that until more is known about the precise way in which lipids and proteins are bound together in biological systems, little further progress can be made towards solving such outstanding problems as the structure of cell membranes. This lack of information is due to the fact that most lipoproteins are fairly loose associations, rather unstable, and exchange their lipid moieties quite readily with lipid in the environment. Many lipoproteins, too, are attached to membranes where a whole range of lipids and proteins are aggregated together and the picture is very complex. For this reason, much of the work on lipid-protein interactions has been done with 'model systems' — in other words, by studying the interaction of various single lipid species or defined mixtures with a purified protein. Although the results of such experiments can tell us a great deal about certain lipid-protein interactions, they may not *necessarily* tell us much about what happens *in the cell* and very great care must be taken when trying to extrapolate the results to natural systems.

Examples of each type of bonding can be found in naturally occuring lipoproteins.

As we cannot specify *precisely* the type of lipid-protein bonding in any one class of lipoprotein we can best quote the results of a limited number of experiments and from these data infer that the bonding is of a particular type.

When the proteins albumin or haemoglobin are injected into water below a negatively charged monolayer of phosphatidyl glycerol, the protein is adsorbed provided that the bulk pH is below the isoelectric point of the protein — in other words, so that the protein has the opposite charge to that of the film. When the pH is increased to give the protein a charge with the same sign as that of the lipid film, adsorption no longer occurs, suggesting an electrostatic interaction between charged groups on the lipid and protein. Mammalian cytochrome c is a basic protein which will combine only with acidic (negatively charged) phospholipids. The fact that bacterial and plant forms of cytochrome c, which are uncharged, will not form the complexes again argues for electrostatic interactions. Mitochondrial *structural protein* (a very insoluble non-enzymic protein of the mitochondrial membrane) on the other hand forms complexes with all types of lipids and the binding is unaffected by the presence of salts or other agents which might cause a weakening of electrostatic forces. This suggests that hydrophobic bonding is predominant in this case. Certain lipoproteins from human serum are ideal for studying lipid-protein interactions because the lipid can be removed with ethanol-ether mixtures, giving rise to a water soluble protein. A lipoprotein, having many of the properties of the original lipoprotein, can be reconstructed by adding the two components together in stoichiometric proportions. When the surface charge of the 'delipidized' protein is altered by chemical modification of certain polar groups (such as acetylation, succinylation, reduction or alkylation), the same stoichiometric proportion of lipid is till bound and the resulting lipoprotein again has many of the properties of the natural compound. *Hydrophobic* interactions again seem to be the major contribution to lipid-protein bonding. For our final example we will turn to the erythrocyte membrane as the site of our interactions and NMR as the analytical tool. Erythrocyte *ghosts* (membranes freed from haemoglobin) give a weak resonance signal for the hydrocarbon protons, suggesting that the apolar chains are 'frozen', probably by interactions with each other or the protein. When the membrane is treated with a detergent, the signal from the hydrocarbon chain region becomes strong, an observation which has been interpreted as due to a breakdown of hydrophobic interactions when the membrane is disrupted by the detergent. The NMR method is at present being exploited to great effect by D. Chapman and his colleagues.

The foregoing paragraphs have shown that the binding of lipids to proteins in nature depends on the attraction of charged groups or of much weaker forces of the *Van der Waals* type. These latter are only really effective if the surface area of contact can be very large or at least the molecules can form multiple points of contact. The shape of the molecules is likely to be very important too. These are the very factors which are thought to govern the binding of enzymes to their substrates (or inhibitors). Making full use of this analogy, the British biochemist R.M.C. Dawson, working in Cambridge, has studied the reactions of phospholipases with their substrates as examples of lipid-protein interactions. Some of the results were described in chapter 4.

It has been implied in some of the examples used in this section that there are two broad types of naturally occurring lipoproteins; first, those which exist as 'discrete' macromolecules (although often ill-defined); these we may call the *soluble types*. Second, those which have become aggregated into a heterogeneous two dimensional structure which we recognize as a membrane. We shall call these *membrane types* and the following two sections will describe in turn some of the properties of the two types.

4. NATURALLY OCCURRING LIPOPROTEINS —'SOLUBLE TYPES'

'Soluble' lipoproteins are usually classified according to their densities because the proportion of lipid to protein in the macromolecule varies widely and this ratio profoundly affects the density. An alternative classification might be based on structure. For example, some proteins might be expected to combine with individual lipid molecules at specific sites, while others could form complexes with whole lipid aggregates or micelles.

Unesterified fatty acids are transported as complexes with albumin.

In fact, combination of individual lipid molecules with specific sites on the protein is relatively rare but occurs in the case of albumin and fatty acids. Serum albumin has three classes of sites which bind 2,5 and 20 molecules of fatty acid respectively. These sites also have a decreasing affinity for the fatty acids in the above order and there are striking differences between the binding constants for different fatty acids at each site. The function of this complex is to transport non-esterified fatty acids in the blood stream to where they may be required for complex lipid synthesis or as an energy source (see chapters 2 and 3). The complex is completely water-soluble and is often used as a tool for introducing fatty acids as substrates into an experimental system (either *in vivo* or *in vitro*) to overcome problems of insolubility in water. The albumin complex is an example of the extreme end of the lipoprotein scale with a high protein and low lipid content. Complexes which are also relatively rich in protein but have the solubility properties of a lipid can be isolated from brain by extraction with chloroform-methanol mixtures. Because of their insolubility in water and solubility in organic solvents they have been called *proteolipids*. They can be purified and shown to contain definite reproducible proportions of protein (50%), cerebroside (20%) and phospholipid

Table. 6.1. Classification and Composition of Human Serum Lipoproteins.

Class	Density range	S_f Range*	‡% protein	% neutral lipids triglyceride	others†	% phospholipid
Chylomicrons	0·94	400	2	83	8	7
VLDL	0·94 — 1·00	20 — 400	9	50	22	18
LDL	1·00 — 1·063	0 — 20	21	11	46	22
HDL	1·063 — 1·20	0 — 6	50	8	20	22

* S_f for LDL determined in NaCl, density 1·063 and for HDL in NaBr, density 1·20.

† Free and esterified cholesterol, non-esterified fatty acids and carotenoids.

‡ These figures neglect the small amounts of carbohydrates often found in lipoproteins (see text).

(30%) and to constitute 2–2½% of brain white matter. Little else is known about them, however, and their function is still obscure.

Most 'soluble' lipoproteins circulate in blood serum and are classified according to density.

The majority of soluble lipoproteins are found in serum and in this case their function is much more obvious, for they represent a convenient way in which a whole variety of water-insoluble compounds can be transported in soluble form between different sites. The remainder of this section will describe their isolation, properties and biosynthesis. Serum lipoprotein classes have been prepared by precipitation with salts, such as ammonium sulphate, with polyanions such as dextran sulphate and by fractional precipitation with ethanol-water mixtures. By far the most general method, however, has been to isolate them according to their densities, by flotation in the ultracentrifuge. Two general methods have evolved. In the

first, a salt solution of fixed density is used – usually sodium chloride of relative density 1·063 at 26°. The rate of flotation (expressed by an S_f value) then depends on the difference in density between the medium and the lipoprotein molecules and the weight and shape of the molecule or particle. In the other method, a density gradient is established by layering solutions of different densities upon each other in the centrifuge tube. During the 18–20 hr normally employed for a centrifugation, the components tend to migrate to the region where the density of the medium corresponds with their own. The technique of flotation by ultracentrifugation enables a classification to be made according to density. Table 6.1 shows the lipoprotein components of human serum. There are four main classes, beginning with *chylomicrons* at the lower end of the density scale, very low density lipoproteins (VLDL), low-density lipoproteins (LDL) and high density lipoproteins (HDL). The range of 'molecular' weight and size is enormous and it should be emphasized that the four classes themselves are not homogeneous, having among

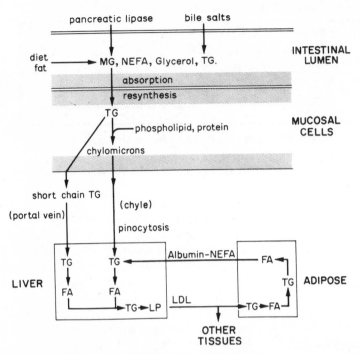

Fig. 6.2. Transport of lipids between the tissues.

a single class, species which may differ in both peptide and lipid moieties.

Chylomicrons have a large triglyceride to protein ratio and are the means by which absorbed fat is transported to the liver.

Chylomicrons contain the largest proportion of lipid and smallest proportion of protein and differ from other serum lipoproteins in being water-insoluble. As you can see in Table 6.1, the lipid is almost entirely triglyceride, for the function of these particles is to transport triglyceride after its resynthesis from absorbed components in the cells of the intestinal

mucosa to the fat depots or the liver (see chapter 3 and Fig. 6.2). Effectively, they may be thought of as small droplets of triglyceride stabilized in the aqueous medium by an 'outer shell' of amphipathic molecules – protein and phospholipids. The largest particles have a diameter of about 4,000 Å and a 'molecular weight' of almost 10^{11}. There seems to be at least three protein components of chylomicrons, one of which is the same as the HDL component. The evidence suggests that the different peptide moieties may be synthesized in different tissues. Most of the chylomicron triglyceride is synthesized by the intestinal mucosal cells and it is probable that the phospholipid and protein components are synthesized there too.

Low density and high density lipoproteins have distinctly different physical properties, structure and metabolism.

As we progress up the density scale, the proportion of lipid decreases while the mean diameter of the particles and 'molecular weight' decreases. LDL and VLDL have diameters of 200–700 Å and contain 8–21% protein (Table 6.1). A complication arises when considering composition because most serum lipo-proteins seem to contain a very small amount (up to 3%) of carbohydrate. In older literature you may come across references to 'α- or β-lipoproteins'. This refers to a classification according to electrophoretic mobility. LDL tend to migrate with the β-globulins during electro-phoresis while HDL tend to migrate with the α-globulins. LDL are water-soluble but the *apoprotein* or 'delipidized' protein is insoluble. In contrast, both the complete HDL and its 'delipidized' protein are soluble in water. Very little can be said about the structure of LDL or HDL except that HDL are thought to have several peptide *subunits* cemented together by lipids. The structure of this class of lipoproteins is currently being studied by A. Scanu and his co-workers in Chicago, by methods involving removal of lipids by solvents and subsequent recombination of lipid and protein. In some cases, notably a HDL from egg yolk named *lipovitellin*, the lipid cannot be completely removed without digesting part of the protein with proteolytic enzymes. This indicates that some of the lipid is 'buried' in the interior. Recent NMR studies indicate that even this buried lipid has considerable mobility.

The major site of both LDL and HDL biosynthesis is in the liver.

How are serum lipoproteins made, and where in the body are they produced? Are the lipid and protein components attached at the site of synthesis or are fully formed lipids attached when required to an *apoprotein* circulating in the plasma? Can one lipoprotein class be converted into another by addition or subtraction of lipids?

The research needed to answer some of these questions is still in the early stages, but the answers will help in our understanding not only of the function of lipoproteins but of the diseases due to abnormalities in their metabolism – a subject which will be mentioned at the end of this section.

A large proportion of lipoprotein production is quite certainly in the liver. There is very little synthesis of plasma lipoproteins in animals which have been 'hepatectomized'; (hepatectomy is surgical removal of the liver). The American biochemist, J. Marsh, has demonstrated lipoprotein biosynthesis in perfused livers, liver slices and in a microsomal fraction from rat liver. He could isolate and identify his product and show that it was identical to a plasma lipoprotein fraction by an immunochemical reaction. There are several difficulties in studying the biosynthesis of lipoproteins. First of all, many people have established that there is considerable 'exchange' between the lipids of one lipoprotein or lipoprotein class and another. Thus, for example, studies involving labelling the lipid moiety are hazardous because some lipoprotein particles may become labelled simply by exchange rather than by *de novo* synthesis. The picture is complicated even further by the fact that some lipid types exchange more readily than others. Cholesterol esters, for example, hardly exchange at all. The exchange process probably involve collision between particles, partial coalescence of the lipid moieties and migration of lipid from one environment to another, followed by separation of the 'new' particles. Another hazard is in the interpretation of differences between data from experiments performed *in vivo* and *in vitro*. Thus labelled amino acids incorporated into lipoproteins *in vivo* have a higher specific activity in LDL than in HDL, whereas *in vitro* the reverse is the case. Again we must ask: what meaning have experiments *in vitro* and how can we relate the results to the truly physiological state?

The rate of lipoprotein production is strictly controlled by the levels and rate of production of tri- glycerides. These in turn are controlled by nutritional and hormonal factors.

One of the problems currently receiving much attention is whether lipid synthesized in the liver is secreted and subsequently attached to an *apoprotein* circulating in the plasma (an hypothesis put forward by several research groups) or whether the complete lipoprotein is synthesized by the endoplasmic reticulum before being released into the circulation.

In some experiments, the biosynthesis of the protein and lipid moieties in liver have been traced by using two different labels at the same time —[^{14}C] leucine as a precursor for protein and [9,10– ^{3}H] palmitic acid for lipid. Two facts clearly emerged. First, the lipid moiety was synthesized earlier than the protein. This suggests that lipid and protein parts are synthesized separately and conjugated later. Second, the secretion of both newly synthesized lipid and protein from the liver into the plasma stopped very quickly after administration of puromycin, a specific inhibitor of *protein* synthesis. If a preformed circulating apoprotein were quantitatively important then puromycin would have no effect on *lipid* secretion. Is it possible that the supply of lipid controls synthesis of lipoprotein- protein or is the protein produced at a constant rate and acquires more or less lipid according to demand? We know that release of triglycerides from the liver depends not only on the hormonal and nutritional state of the animal but also on the hepatic concentration of glycerol- 3-phosphate and the concentration of fatty acids in the portal vein. Perfusion of an isolated liver with a medium rich in linoleic acid causes a three-fold increase in triglyceride synthesis by the liver *and* a similarly enhanced release of LDL-protein. Although the incorporation of labelled amino acids into LDL is increased under these conditions, their incorporation into general proteins is unaffected. It seems, therefore, that lipid *levels* in the tissues have a pro- found effect on lipoprotein synthesis and

release. The levels of lipids and lipoproteins, as we have mentioned, are
under hormonal control and the adrenal,
thyroid and gonadal hormones all play
a part. Patterns of plasma lipoproteins
also respond to different diets. Restriction
of calories in the form of fat or carbohydrate, or both, tends to lower the
levels of LDL but saturated fatty acids
tend to increase and unsaturated fatty
acids to decrease the levels. HDL are
unaffected by dietary conditions – another
pointer to the marked difference between
lipoprotein classes. Interrelationships
between lipids and lipoproteins of liver,
plasma, intestines and adipose tissue
are shown in Fig. 6.2.

*The enzyme lipoprotein lipase
hydrolyses lipoprotein triglyceride and is involved in
maintaining normal lipoprotein
levels.*

We have considered synthesis, but
equally important in any metabolic
system under strict control is catabolism
or breakdown. Little is known about the
degradation of the protein moiety – in
common with the lack of knowledge of
tissue protein breakdown generally. The
enzyme which catalyses the hydrolysis
of triglycerides in LDL or chylomicrons
is lipoprotein lipase, sometimes called
clearing factor lipase because of its
ability to clear 'lipaemic' plasma. This
enzyme cannot be detected in plasma
unless a dose of the polysaccharide
heparin has been adminstered. The
mechanism of the 'activation' is not
understood, but the enzyme is thought
to be associated with adipose cells and
released into the capillaries by the drug.

Defects in Lipoprotein metabolism

*The fine balance of lipoprotein
metabolism is easily disturbed
by diseases of various tissues.*

As we have seen, a great number of
factors including diet, hormones, and
lipase activity operate to maintain precise
levels of lipoproteins in the serum of
'normal' animals. When one or more of
the tissues is affected by certain types
of disease, this can be reflected in a
marked change in lipoprotein levels.
Certain 'experimental diseases' can be
useful tools in the study of lipoprotein
metabolism. Kidney disease (nephrosis),
for example, results in *hyperlipaemia* or
increased levels of LDL. HDL are not
affected. One possible explanation is
that the disease causes abnormal loss of
protein from the kidney, resulting in
protein deficiency. The liver tries to
compensate by increasing its supply of
proteins, including lipoproteins. HDL are
small enough to be excreted again by the
diseased kidney but the larger LDL
accumulate. *Diabetes mellitus* and
coronary artery disease are also associated
with an increase in LDL levels but the
metabolic block is unknown. Liver
disease, which can be caused by certain
toxic chemicals such as carbon tetra-
chloride, is characterized by a large
accumulation of liver triglycerides and
is therefore called *fatty liver.* The
primary action of the toxic chemical is
believed to be disruption of the protein
synthesizing mechanism of the endoplasmic reticulum. Lipoprotein-protein
production is thereby reduced and triglycerides, which would otherwise have
been excreted into the plasma as LDL,
accumulate in the liver.

Finally, there are a number of congenital disorders — termed *hyper-* or *hypo*lipoproteinaemias according to whether the level of serum lipoprotein is abnormally high or low.

(a) A-β-Lipoproteinaemia (familial low density lipoprotein deficiency)

The *clinical* signs of this disease are neuromuscular disturbances, retinal changes, morphological abnormalities of the red cells and steatorrhea (production of bulky, excessively fatty stools). *Biochemically* there is a deficiency or lowering of serum β-lipoprotein, and lack of ability to form chylomicrons after a fatty meal. All recent evidence favours a defect in synthesis of the specific protein moiety of the β-lipoproteins and chylomicrons rather than a defect in the mechanism of coupling lipid to protein. The primary defect is probably in the mucosal cells of the intestine so that long chain fatty acids are only poorly absorbed, and not transmitted to the lymphatic tissue. Absorption of short and medium chain fatty acids (i.e. below C_{10}) is less affected since these are carried by the blood in the portal vein.

Unlike the sphingolipidoses which we have already described in chapter 5, A-β-lipoproteinaemia can be improved by symptomatic measures such as limitations of fat intake (which reduces the steatorrhea), by the use of medium chain triglycerides in the diet, and by decreasing animal fat and substituting unsaturated vegetable oils such as cotton seed or corn oils (which results in production of red cells with a normal morphology).

(b) Tangier disease (familial high density lipoprotein deficiency)

This rare disease was first described by Frederickson's group at the National Institutes of Health, U.S.A., and was named after the island of Tangier in Chesepeake Bay from which the first patient came. The major clinical findings are discolouration and gross enlargement of the tonsils, lipid accumulation (particularly cholesterol esters) in reticulo-endothelial tissues and enlarged spleen, liver and lymph nodes. The familial nature of the disease is shown by the fact that of the close relatives of the first two patients, 70% of the males and 50% of the females had abnormally low plasma high density lipoproteins. The disease is not progressive.

(c) Essential hypercholesterolaemia (familial hypercholesterolaemic xanthomatosis)

This hereditary disorder is *biochemically* defined by its elevation of plasma cholesterol and phospholipid levels caused by increased production of β-lipoproteins. *Clinically* the disease shows often massive accumulation of cholesterol in the skin, tendon etc. as xanthomas. The incidence of this disease is greater than the other lipidoses; a prevalence of 12% was found in 300 subjects in New York. In subjects with the disease, the plasma cholesterol was above 324 μg/100 ml in children and above 340 μg/100 ml in adults as compared with below 220 μg/100 ml in 'normals'. It is thought that the disease in children is due to their being carriers of a double abnormal gene whereas its occurrence in later life may be due to only one abnormal gene. Sudden death due to ischaemic heart disease is more frequent in those families with a high incidence of hypercholesterolaemia. Indeed about 30% of all cases have symptoms of involvement of the cardio-vascular system.

As with A—β—lipoproteinaemia, modification of the diet can provide alleviation of symptoms. The usual treatment is a restriction of fat intake (no more than 25% of total calories) and an increase in the percentage of poly-unsaturated fatty acids. However, this rarely produces a lasting fall in plasma cholesterol levels.

The increased β-lipoprotein level is probably due to an increase in the protein moiety rather than a shift in cholesterol metabolism, though an altered phospho-lipid metabolism is possible.

---

5. NATURALLY OCCURRING LIPOPROTEINS — 'MEMBRANE TYPES'

Although in the past there have been theories to explain the concentration differences inside and outside cells without invoking a physical barrier, and even in modern literature one can find such ideas, the idea of a *membrane* which preserves the integrity of the cell and regulates transport processes is generally accepted. The presence of lipids in cellular membranes was first proposed in the nineteenth century to account for the observed relationship between lipid solubility and the velocity of penetration of compounds into cells. Now that techniques for isolating membrane material free from soluble cytoplasmic components are readily available in many cases, the presence of lipid is in no doubt and the complete chemical composition of many membranes can be determined. This is an important preliminary in trying to determine the structure of the membrane— one of the greatest preoccupations of modern biology.

A detailed knowledge of the lipid and protein composition of different membranes is an essential prerequisite for studying their architecture

Certain cellular structures are easier to obtain in an homogeneous state than others and therefore have received more attention. For example, the erythrocyte contains no subcellular structures or cytoplasmic membranes and its membrane or 'ghost' can easily be washed free from haemoglobin and other soluble contaminants. The myelin membranes of nervous tissue, too, are easy to isolate and have been a favourite source of chemical and structural information. Unfortunately, these materials have too often been used as models to make wide generalizations about membrane structure, whereas a brief look at their chemical composition alone will show us that there is no such thing as a 'typical' membrane. It is not our purpose in this book to give exhaustive data on

Table 6.2. Lipid composition of different membranes

	Chloroplast (spinach)	Protoplast (B. Megaterium)	(% total lipid) Membrane Mitochondrion (rat)	Erythrocyte (rat)	Myelin (rat)
Lipid: protein	1:1	1:3	1:3	1:3	3:1
Phospholipid	12	48	90	61	41
PC	tr	0	40	34	12
PE+PI+PS	tr	19	41	11	26
PG	12	26	–	–	–
CL	–	3	7	–	–
SPH	–	–	2	16	3
Glycolipid	80	52	–	11	42
MGDG	41	–			
DGDG	23	–			
SL	16	–			
Sterol, sterol ester	tr	0	tr	28	17
Glycerides	–	–	–	10	–
Pigments	8	–	–	–	–

the lipid composition of tissues and membranes. Such information is readily obtainable elsewhere (see bibliography at the end of the chapter); we merely wish to make a few pertinent comparisons to allow us to take a critical view of the theories on membrane structure which will be discussed later.

In Table 6.2 we can compare the lipid composition of a few different types of membranes. One notable point is the composition by weight of lipid in the membrane. All membranes have more than 25% by weight of lipid, but myelin is outstanding in having as much as 75%. Neutral lipids such as triglycerides have little part in membrane structure. The ratio of steroids to phospholipids or glycolipids is low in the intra-cellular membranes like chloroplast lamellae or mitochondria but relatively higher in the membranes which surround cells such as the erythrocyte or plasma membrane, or in the specialized structure, myelin. The exception here is the bacterial protoplast membrane where steroids are not found at all. As far as polar lipids are concerned, note the preponderance of glycolipids in the plant chloroplast membrane and phospho-lipids in the animal membranes.

The lipids of a given type of membrane, such as that of the red cell, may vary markedly among different species. In Table 6.3 note particularly the steady decrease in phosphatidyl choline content in rat through to sheep. At the same time, there is a corresponding increase in the sphingomyelin content, so that in the ruminant this is the major choline-con-taining lipid. The net effect is a relative constancy of the total choline-containing

Table 6.3. Phospholipid composition of erythrocyte membranes of different species

| Species | % of total phospholipids. | | |
	PC	SPH	PE+PS+PI
Sheep	1	63	36
Ox	7	61	32
Pig	29	36	35
Human	39	37	24
Rabbit	44	29	27
Rat	56	26	18

lipids. Professor van Deenen's group in the Netherlands have correlated these compositional changes directly with differences in permeability to certain molecules.

The discussion so far has considered what are effectively changes in the polar headgroups of the membrane lipids. Our earlier discussion of the physical properties of lipids indicated that hydrophobic interactions will be no less important. Are the fatty acid patterns of all cell membranes similar or can we find an association of particular fatty acids for certain membranes or certain lipids? Table 6.4 pinpoints certain features of the fatty acid distribution in the lipids of chloroplast lamellae and liver mitochondria which help to answer this question. The immediately outstanding feature is the preponderance of unsaturated acids, particularly in the chloroplast. Trans-3-hexadecenoic acid, a fatty acid peculiar to photosynthetic tissue, is esterified exclusively in one chloroplast lipid – phosphatidyl glycerol. Cardio-lipin (diphosphatidyl glycerol), a lipid which accumulates almost exclusively in the mitochondria of mammalian cells, has a preference for linoleic acid.

Table 6.4. Fatty acid composition of different lipids within a single membrane

	Chloroplast (Spinach)							
	16:0	$t\Delta$ 3–16:1	16:3	18:0	18:1	18:2	18:3	others
MGDG	—	—	25	—	—	2	72	1
DGDG	3	—	5	—	2	2	87	1
SL	39	—	—	—	—	6	52	3
PG	11	32	2	—	2	4	47	2
PI	34	—	3	2	7	15	27	12

	Mitochondria (Rat liver)						
	16:0	18:0	18:1	18:2	20:4	22:6	others
PC	19·7	18·7	12·2	20·0	19·5	3·4	6·5
PE + PS	20·2	19·5	10·4	16·3	21·0	8·7	3·9
PI	12·1	32·0	6·8	16·2	21·5	3·4	8·0
CL	39	1·4	12·8	74·0	1·6	0·4	5·9

Much less is known about membrane proteins than about the lipid components.

While a great mass of data has been published on the lipid composition of membranes (mainly mammalian), knowledge of the nature, composition and structure of the proteins involved is rather scanty. Many of the proteins are enzymes, for many biochemical reactions take place in the membrane; they need an organized structure of this type to be able to function. Examples are the various *permeases* needed to transport certain molecules, such as sugars, across the cell membrane; or the electron transport system of the mitochondria, the complete functioning of which requires a fully integrated membrane structure. Many research workers believe that, in addition, there are *structural proteins* which have no enzymic activity but which serve as the backbone of the membrane. All structural proteins isolated so far, from whatever source, are similar in composition and properties. They have a large proportion of hydrophobic amino acids (alanine, leucine, phenylalanine, valine), conspicuously lack —SH groups and have a preponderance of acidic rather than basic amino acids.

A brief look at the history of cell membrane theories.

It is sometimes illuminating, as in certain works of fiction, to begin a historical survey at the end! In recent years, several research groups have been able to isolate lipoprotein particles by fragmenting membranes with detergents or by mechanical means such as ultrasonication. Under appropriate conditions these particles appear to reaggregate spontaneously into structures which have all the appearance — microscopically and biochemically of natural membranes. By examination of the conditions needed for such aggregation and by studying the kinetics of the process we may hopefully begin to emerge from the confusion which has prevailed in this field for many a long year. In any case membranes begin to gain perspective as

1. Gorter and Grendel's bimolecular lipid membrane. Circles represent the polar ends of the molecule while the bars represent the long chain hydrocarbon moieties.

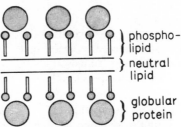

) phospho-lipid
) neutral lipid
) globular protein

2. The Danielli-Davson Model. Two monolayers of phospholipid are on either side of a layer of neutral lipid of unspecified thickness. The polar ends of the phospholipid molecules are associated with a monolayer of globular protein.

lipid   protein   lipid

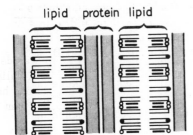

3. Myelin membrane according to Finean. Lipid bimolecular layers contain phospholipid (⇒) glycolipid (⇒•) and cholesterol (—)
   An example of a Robertson "Unit Membrane"

4. Membrane lipoprotein subunit of Benson Hydrophobic lipid chains are associated with the hydrophobic interior of the protein.

5. Schematic representation of the inner membrane of the mitochondrion according to Green. The base pieces (⬭) are the lipoprotein subunits of the continuous membrane, and line up end to end provided lipid is present. The stalks (우) are detachable and part of the superstructure of the membrane.

Fig. 6.3. Some proposed models for biological membranes.

lipoproteins which have the ability to organize themselves in a two dimensional structure having a specialized function in contrast to the free-floating 'transport lipoproteins' which we have previously discussed.

The presence of a barrier at the periphery of a cell could long ago be inferred, but the development of more and more powerful microscopial techniques has opened up possibilities for investigating its structure and also revealed the presence of intracellular (cytoplasmic) membranes.

The Bimolecular Lipid Leaflet model

In 1925, the Dutch workers, E. Gorter and F. Grendel extracted the lipids from erythrocytes and calculated the area occupied by the lipid from a known number of cells when it was spread as a monolayer on a *Langmuir Trough*. There was sufficient, they claimed, to surround a red cell in a layer two molecules thick (Fig. 6.3). Although their techniques can be criticized, the idea of a *bimolecular lipid leaflet* has survived in several theories up to the present day. J.F. Danielli and H. Davson, unaware of Gorter and Grendel's paper, proposed a similar model in 1935 but realized that the measured surface tension of a membrane was too low to be accounted for by lipid and proposed that a layer of protein was present at each surface. With the advent of electron microscopy came pictures of membranes which seemed to support the Danielli model. For the purposes of electron microscopy, the specimen has to

be dehydrated, stained (usually with osmium tetroxide or potassium permanganate — so-called *positive staining*) and embedded in a plastic material such as an epoxy resin. Thin sections are then cut and examined under the microscope. Much of the interpretation of the structures observed depends on how much shrinkage occurs during dehydration of the sample, and what regions of the membrane take up the stain: it is now generally accepted that osmium tetroxide accumulates at the polar regions of lipid and protein. In just about every membrane examined in this way, a *triple-layered* structure (two dark lines on either side of a light band, see Fig. 6.4a), was seen on the micrographs and on this basis, the American microscopist, J.D. Robertson put forward the *unit membrane hypothesis* which held that every membrane had a basic structure consisting of a bimolecular lipid leaflet sandwiched between a layer of protein on one side and *glycoprotein* on the other. Further support for the hypothesis came from X-ray studies on myelin by the British biophysical chemist J.B. Finean. Myelin is an ideal material for this purpose: it can be easily isolated in a pure state, has a simple composition and regular repeating features which provide good diffraction patterns. Most of the evidence, in fact, supporting the unit membrane hypothesis comes from studies with myelin (Fig. 6.3). To sum up the evidence in favour of this model: (1) the physical-chemical studies which we discussed earlier show that lipids can take up a lamellar configuration; (2) observation of a ubiquitous three-layered pattern in fixed, stained tissue sections under the electron microscope are in keeping with this idea; (3) X-ray diffraction patterns of myelin are also in accord with the model. There are several limitations, however. Examination of Fig. 6.3 will show that electrostatic bonding of lipid to protein is implicit in the unit membrane model. Both membrane lipids and membrane proteins are on the whole rather acidic and the weight of evidence at present suggests that *hydrophobic* bonding is the more important. Furthermore, extraction of more than 90% of the lipid from the mitochondria results in a membrane preparation which has a 'normal' appearance under the electron microscope; this fact has presented something of a puzzle for those looking for a major role for lipids in membranes.

In Tables 6.1 to 6.3 we have tried to indicate that from the point of view of composition at least, there is no such thing as a typical membrane. With such widely differing compositions it is unreasonable to expect a universal structure. Myelin, whose structure lent most support to the unit membrane theory, seems to be the most atypical.

The Globular Sub-unit model.

In recent years, two new techniques in the electron microscopist's armoury have helped to encourage a re-appraisal of membrane structure. One is *negative staining*, in which the sample is not fixed and embedded but dispersed in an aqueous solution of the negative stain (phosphotungstate) and dried down on a support film (Fig. 6.4b). Stain accumulates in the hydrophilic regions. The other method, drastically different from the staining methods and therefore useful to give independent corroboration, is the *freeze-etching* technique (Fig. 6.4c). These approaches have indicated that certain membranes are composed of *globular sub-units*. These have been interpreted as lipoprotein particles with lipid hydrocarbon chains and protein apolar side chains in the centre. We should be aware of a profound change of approach of this model, in that we are now dealing with individual lipoprotein particles in contrast to separate and continuous lipid and protein phases. Mitochondria can be fragmented into *sub-units* — particles having the enzymic activities of part of the electron transport chain — by ultrasonic disruption or by treatment with detergents which loosen the bonds between sub-units. When the detergent's concentration is reduced by dilution or dialysis, the individual sub-units appear to re-aggregate into structures which have all the appearance of membranes under the electron microscope. The 'sub-units theory' gained support not only from new techniques in electron microscopy but from the physical-chemical findings discussed earlier that lipids can adopt phases other than lamellar.

No one model can yet be considered as a true picture of a natural membrane. Transitions between different conformations may occur within a single membrane.

The electron microscopist F. Sjöstrand argues for at least two broad types of membranes on grounds of dimensions, ultrastructure and function. The first

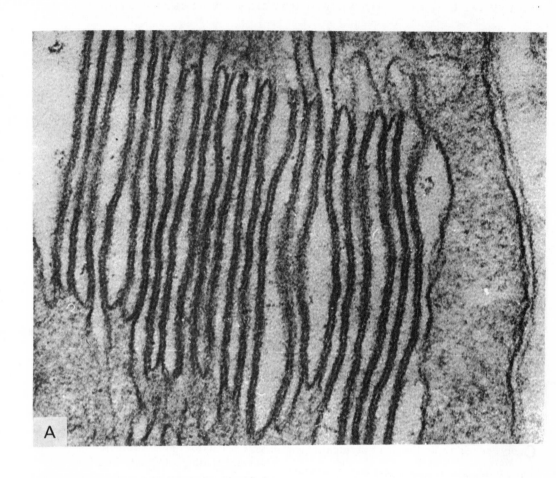

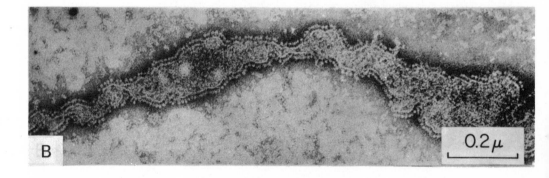

Fig. 6.4. Examples of some different techniques of electron microscopy used in examining membrane preparations.

A. An electron micrograph of the chloroplast lamellae of a green narcissus petal. The specimen was fixed in glutaraldehyde osmium, embedded in epon and post-stained with lead citrate. This is an example of the 'positive staining' technique, and clearly illustrates the typical 'unit membrane' feature of two dark lines separated by a light band, Magnification: × 142,500

B. An electron micrograph of beef heart mitochondrial membranes. This is an example of the 'negative staining' technique, and illustrates the regular array of globular lipoprotein particles (known as 'elementary particles') attached to the membrane. Some detached particles can be seen. Magnification: × 93,000

C. An electron micrograph of a 2% dispersion of dioleoyl phosphatidyl choline in water. This is an example of the 'freeze-etching' technique and illustrates the way in which phosphatidyl choline molecules take up a lamellar configuration when dispersed in water. Magnification: × 98,000

197

type would be exemplified by plasma membranes. They are 90–100 Å thick and almost certainly have the bimolecular lipid leaflet structure. These membranes shield the interior of the cell from the outside and impose restrictions on the penetration of ions and molecules into, and out of, the cell. As far as structure is concerned, at least, myelin would belong to this group. To the second type belong the cytoplasmic membranes such as the mitochondria or the endoplasmic reticulum with a thickness of 50–70 Å. They can be considered primarily as 'metabolic' and as the location of the cell's multienzyme systems. Of course, the distinction cannot be pushed to extremes, for the plasma membrane has considerable metabolic activity, while the mitochondrion has to regulate the flow of metabolites between the cytoplasm and the intramitochondrial space.

We should emphasize that the models described here represent two rather extreme views of membrane structure. In between, there are many different possibilities and equilibria between different types of structures should also be considered. Volumes have been written on the subject and the student is encouraged to use the bibliography at the end of the chapter.

The role of lipid is still uncertain. Many views have been, and are still being, put forward.

If, as was stated earlier, the mitochondrial lipid can be extracted without altering the appearance of the membrane, what is its role, and how can it ultimately be bound up with structure and function as we have implied throughout this chapter? First of all, it should be emphasized that the statement that the membrane was unchanged was based solely on electron microscopic evidence and that *all* electron microscopic interpretations should be examined with extreme care. Second, although the apparent structure is unaltered, the ability of the membrane to perform its proper functions, namely integrated enzymic reactions such as electron transport, is completely destroyed. This function can be restored by 'adding back' the lipids, in a micellar form, to lipid-free 'membranes' It is important that the lipids are added as very small micelles, in other words as a water-clear, and not a turbid, suspension.

It has been suggested that one of the roles of lipid in the mitochondria is to provide a medium of low dielectric constant, in which electron transport can take place and in which the high energy bonds of ATP, which it is the prime function of the mitochondria to manufacture, can survive. More recently, the research team at the Enzyme Institute have put forward the view that the primary role of lipid is to bind to specific faces of the *unsymmetrical* protein sub-units, so that the resulting lipoprotein particles can combine only in a certain pattern to form a two-dimensional sheet, which ultimately becomes a membrane. When lipid is extracted from the *sub-units* the delipidized particles can combine in a three-dimensional aggregate in which the majority of the enzyme sites are buried in the interior. These lipid-free aggregates are therefore enzymically inactive. Removal of lipid from an already formed membrane, they think, does not disrupt the bonding already present between adjacent protein sub-units.

*Several purified enzyme proteins
need to be combined with lipids
in order to have full enzymic
activity.*

Not only whole membranes but individual
segments of the electron transport chain
and even individual purified enzymes
themselves may have a dependence on
lipid in the same way as described for
whole mitochondria. In the case of
enzymes, one might ask whether lipids
could act as coenzymes? This seems
unlikely in view of the relatively large
proportion of lipid to protein present in
enzymes of this type. In most cases,
too, the lipids required to restore
activity to extracted preparations are
unspecific. A notable exception is the
mitochondrial enzyme, *D-3-hydroxybutyrate
dehydrogenase*, which has a specific
requirement for phosphatidyl choline. No
other pyridinoproteins of this type require
lipids but they all possess a hydrophobic
region which may be required for interaction of the substrate with NAD.
Phosphatidyl choline might fulfil this
role in the case of *D-3-hydroxybutyrate
dehydrogenase*. On the other hand, the
microsomal enzyme *glucose-6-phosphatase*,
although containing a large amount of
phosphatidyl choline, is not reactivated
readily by adding back this phospholipid,
although phosphatidyl ethanolamine is
is very effective. The lipid requirement
of this enzyme was demonstrated, not by
solvent extraction, but by phospholipase
degradation. When this method is used,
adequate controls have to be used to
ensure that the hydrolysis products —
often powerful detergents — are not themselves the cause of diminished enzyme
activity.

Many other enzymes have recently been
shown to need lipid for activity. In some
cases, such as *cytochrome oxidase*, there
is still controversy about whether the
purified preparation contains lipid and
therefore whether lipid is obligatory for
enzyme activity. In the case of *succinate
dehydrogenase* from mammals, it seems
to depend on what segment of the
respiratory chain is being measured;
when a dye is used as electron acceptor
at the level of the dehydrogenase molecule,
the activity is not affected by phospholipase treatment. In contrast, lipid is
more directly implicated in the yeast
enzyme, where succinyl-phosphatidyl
choline is an active intermediate in the
oxidation of succinate.

The role of lipid in membranes and
enzyme systems is certainly not the same
in all instances and in most cases is
ill-defined. Most of the explanations for
a lipid role have more than an element
of speculation about them, and it will
be a year or two before we can make a
rational judgement on the whole subject.

6. LIPOPOLYSACCHARIDES AND BACTERIAL CELL WALLS

Not all of the 'bound lipid' of living cells
is complexed with protein. While the predominant material of all animal membranes
and plant intracellular membranes is
lipoprotein, in bacterial cell walls lipid
is bound quite extensively with complex
polysaccharides as *lipopolysaccharide.*

The lipids of gram-positive
bacteria are located mainly in
the 'protoplast membrane'.

Most gram-positive bacteria have a wall
whose structural material is a glycopeptide, together with complex polysaccharides
and polymeric material called *teichoic
acids*. There is little or no lipid, virtually
all of the lipid of these organisms being
found in the *protoplast membrane* — the
lipoprotein structure which remains after
removing the outer wall with digestive
enzymes such as *lysozyme*. However, as
in most things, there are notable exceptions,
chiefly the mycobacteria, which may
contain up to 60% of the wall as lipid!
Although these may contain the usual
phospholipids described in chapter 4,
the lipids most characteristic of these
organisms are the waxes. The waxes
contain about 50% of mycolic acid and
50% of a water-soluble glycopeptide.

Gram negative bacteria have a
large amount of lipid in the wall,
mainly associated with the
antigenic lipopolysaccharide.

In gram negative bacteria, the wall is far
more complex and contains glycopeptide,
lipopolysaccharide, phospholipid and
protein. There is not such a clear
distinction between the wall and the
membrane as in gram positive bacteria.
Up to 20% of the wall contents may be
lipid. Only a small amount is 'extractable'
with the usual solvents; the rest is
extractable only after acid hydrolysis,
suggesting that covalent bonding may
be important here in contrast to the bonding
in lipoproteins. However, little can be
said at this stage about the structure of

the lipids in the cell wall because of its
extreme complexity, and the quite large
variations among the many thousand
species of gram negative bacteria.

Lipids are essential components
of the enzyme system which
synthesizes cell wall lipo-
polysaccharide.

The lipopolysaccharide component is
responsible for the antigenic properties
of the cell and is interesting in that its
biosynthesis requires the participation
of a phospholipid. In Fig. 6.5a is shown
a diagramatic representation of the
structure of the cell wall as suggested
by O. Westphal in Germany, and Fig. 6.5b
indicates the partial structure of the
lipopolysaccharide. Two of the steps in
its biosynthesis involve the transfer,
first of glucose, then of galactose on to
the partly synthesized 'acceptor'. The
steps are catalysed by *glycosyl transfera*
(reactions 1 and 2 in Fig. 6.5b), and the
active species which are transferred by
these enzymes are the UDP-sugars. The
way in which these reactions were studie
by Drs Rothfield, Horecker and their
colleagues in the U.S.A. is a good
illustration of the growing use of bacteria
mutants in modern biochemistry. Mutants
were found which lacked the ability to
transfer the glucose and galactose
moieties respectively, and therefore whose
lipopolysaccharides lacked the portion
either from glucose onward or from galact
onward (Fig. 6.5b). These 'deficient'
lipopolysaccharides were isolated from
the respective mutants and used in a
cell-free assay system as *acceptors* for
the transferases. In each case there was
no reaction unless a phospholipid — the

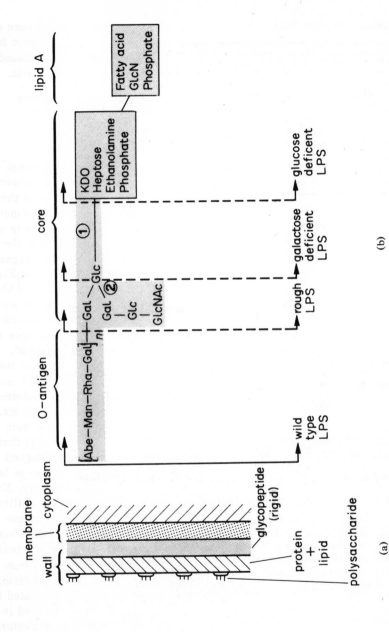

Fig. 6.5. (a) structure of the cell wall (b) structure of the lipopolysaccharide. Abe = abequose; Man = mannose; Rha = rhamnose; Gal = galactose; Glc = glucose; GlcN = glucosamine; GlcNAc = N-acetyl-glucosamine; KDO = ketodeoxyoctonic acid; LPS = lipopolysaccharide.

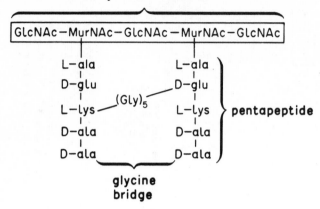

(a)

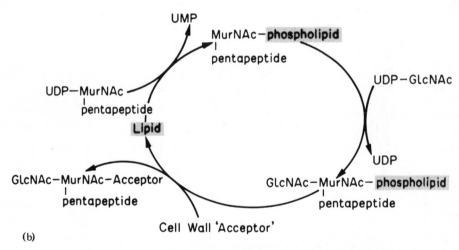

(b)

Fig. 6.6. (a) structure of the glycopeptide. (b) biosynthesis of the glycopeptide.
GlcNAc = N-acetyl glucosamine; Mur-NAc = N-acetyl muramic acid
(muramic acid is 3-O-lactyl glucosamine); ala = alanine; glu = glutamic
acid; lys = lysine; UDP = uridine diphosphate; UMP = uridine
monophosphate.

most effective of which was phosphatidyl ethanolamine containing unsaturated or cyclopropane fatty acids — was introduced in a specific way into the reaction mixture. The lipid and lipopolysaccharide acceptor had to be mixed by heating followed by slow cooling prior to addition of the transfer enzyme. The binary complex (lipid-LPS) then forms a further (ternary) complex by specifically binding the particular transferase corresponding to the 'deficient' lipopolysaccharide. In other words, the phospholipid-galactose-deficient-lipopolysaccharide specifically binds galactosyl, but not glucosyl, transferase.

peptide; this is the material whose synthesis is inhibited by the penicillins. It consists of a polysaccharide backbone, with cross-linked peptide side chains (Fig. 6.6a). *N*-acetyl muramic acid-pentapeptide units are first synthesized and then linked via their UDP derivatives to *N*-acetyl glucosamine to build up the backbone. Before reaction with UDP-*N*-acetyl glucosamine can occur, however, the *N*-acetyl-muramyl-pentapeptide must first be transformed into a lipid-containing complex. The reactions are shown in Fig. 6.6b. The lipid is a derivative of a C_{55} isoprenoid alcohol.

As will be evident from this brief

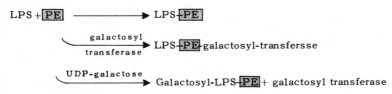

Nothing is yet known about the way in which phosphatidyl ethanolamine helps bind the enzyme and lipopolysaccharide together.

A phospholipid intermediate is involved during the building up of the cell wall glycopeptide backbone.

One of the chief polymers which gives the cell wall its rigidity is the *glyco-*

account, the bacterial cell wall lipids, in contrast to the animal lipids, are ill-characterized mainly due to the complexity of the wall and the bewildering variety of bacterial species. The bacteria, however, offer considerable scope for studying the genetic control of the pathways of lipid metabolism, an area of study which hitherto has not impinged upon the field of lipids and which we can expect to see opening up in the next few years.

SUMMARY

The chemical groups present in both lipids and proteins are such that weak inter-actions like *Van der Waals* forces or hydrogen bonds are the predominant forces binding these molecules together. In specific cases, more powerful electro-static forces may be important, but covalent bonding is rare. The resulting complex molecules — lipoproteins — exist either as discrete particles which circulate in body fluids such as serum or lymph, or as components of two-dimensional macromolecular structures known as biological membranes. The term *serum lipoproteins* covers a wide range of aggregates of differing lipid content. Their precise structure is still uncertain but their main function is the transport of lipids between body tissues. The arrangement of lipids and proteins in cell membranes is still a question of debate. Most of the data relating to their structure derives from X-ray diffraction and electron microscopic studies but today, newer techniques such as NMR are being exploited. Two major models have evolved since 1925 to describe membrane structure and they represent extreme cases. The *bimolecular lipid leaflet* model pictures membranes as bimolecular layers of lipid molecules sandwiched between sheets of proteins. Electrostatic bonding between lipid and protein moieties is implicit in this model. At the other extreme, we have a model based on repeating *sub-units* of lipoprotein particles, in which the bonding is mainly hydrophobic. From the point of view of microscopic appearance, chemical composition and function, biological membranes differ widely and it is probable that the true picture lies between these two extremes, with some membranes having a more 'globular' structure, and others having more of the characteristics of the bimolecular leaflet. Transitions between these types cannot be ruled out.

The *structural* role of lipids cannot be divorced from a *functional* one as many enzymic proteins lose their activities when lipid is extracted and regain them when the lipid is recombined. No *single* role can be assigned to lipids and in no case is the function completely understood.

Lipids also occur in combination with carbohydrates as the *lipopolysaccharides* of bacterial cell walls. The biosynthesis of both the lipopolysaccharide (which is responsible for the antigenic properties of the organism) and the glycopeptide (which is the component giving rigidity to the wall) are dependent on the presence of phospholipids.

BIBLIOGRAPHY

Physical-chemical aspects

1. GURD F.R.N. (1960). Association of lipids with proteins, chapter 8 in *Lipide Chemistry*, edited by D.J. Hanahan, John Wiley, New York and London.

2. CHAPMAN D. (1965). *The Structure of Lipids by Spectroscopic and X-ray Techniques*. Methuen, London.

General

3. ANSELL G.B. and HAWTHORNE J.N. (1964). Phospholipids, chapters 7, 8, 10 and 11, Elsevier, Amsterdam.

4. DAWSON R.M.C. (1966). The metabolism of animal phospholipids and their turnover in cell membranes, in *Essays in Biochemistry*, Vol. 2, page 69, edited by P.N. Campbell and G.D. Greville (The Biochemical Society), Academic Press, London and New York.

Serum Lipoproteins

5. SCANU A.M. (1965). Factors affecting lipoprotein metabolism, *Advances in Lipid Research* 3, 63, edited by R. Paoletti and D. Kritchevsky, Academic Press, New York and London (see also Chapter 9 of reference 1).

Biological Membranes and Bacterial Cell Walls

6. VAN DEENEN L.L.M. (1965). Phospholipids and biomembranes, *Progress in the Chemistry of Fats and other Lipids*, Vol. VIII, part (i). Edited by R.T. Holman, Pergamon Press, Oxford.

7. CHAPMAN D. (ed.) (1968). *Biological Membranes, Physical Fact and Function*, Academic Press, London and New York.

8. ROTHFIELD L. and FINKELSTEIN A. (1968). *Membrane biochemistry, Ann. Rev. Biochem.* 37, 463.

9. LENNARZ W.J. (1966). Lipid metabolism in the bacteria, *Advances in Lipid Research* 4, 175, edited by R. Paoletti and D. Kritchevsky, Academic Press, New York and London.

Lipoproteins: General

10. TRIA, E. and SCANU A.M. (ed.) (1969). *Structural and Functional Aspects of Lipoproteins in Living Systems*, Academic Press, London and New York.

Appendix

Brief notes on some recent developments in lipid research which have been published since we prepared this text

---

FATTY ACIDS: UNSATURATED (*Refer to pages 54 and 55*)

NADH: cytochrome b_5 reductase has been implicated as a component of the desaturase multienzyme complex of hen liver endoplasmic reticulum. Treatment of hen liver microsomes with deoxycholate dissociates the complex into two fractions, one of which contains NADH: cytochrome b_5 reductase. Neither fraction alone has desaturase activity, but full activity can be restored by recombination of the two fractions (Holloway, P.W., and Wakil, S.J. (1970) *J. Biol. Chem.* **245**, 1862).

Desaturases present in hen liver, *Chlorella vulgaris*, goat mammary gland and *Torulopsis utilis* which catalyse the formation of monoenoic acids, insert a double bond in the 9, 10-position regardless of the chain length of the saturated substrate. Maximum activity, however, is always with the C_{18} chain. The enzyme evidently 'measures' the double bond position from the carboxyl group which is probably bound to the enzyme surface at a specific site.

Chlorella vulgaris appears to have two enzymes for catalysing the formation of a second double bond. One of these forms $\Delta 9$, 12-dienes from $\Delta 9$-monoenes, the other forms $\omega 6$, 9-dienes from $\omega 9$-monoenes, regardless of chain length. The latter may be the enzyme which utilizes the lipid-bound form of the substrate (p. 54). Monoenes which have the double bond in neither the $\Delta 9$ nor $\omega 9$-positions are not substrates, except 12-octadecenoic acid. This is desaturated by the hen liver destaurase, which normally converts stearic into 9-octadecenoic acid, to form 9, 12-octadecadienoic acid and hence this is an example of the formation of linoleic acid by an animal tissue, if it is presented with an unnatural substrate.

FATTY ACIDS: PROSTAGLANDINS (Refer to page 62)

During 1970, prostaglandins have been successfully used for induction of labour and therapeutic abortion. For abortions, a solution of prostaglandin E_2, at a concentration of $5\,\mu g/ml$, was administered by intravenous infusion at the rate of $5\,\mu g/min$ until abortion was complete (about 15 hours). The method appears, so far, to be safe, with few side effects, and with no further surgical intervention required (Editorial (1970) *Lancet* **1**, 927; Karim, S.M.M. and Filshie, G.M. (1970) *Brit. Med. J.* **3**, 198).

TRIGLYCERIDES: ALTERNATIVE BIOSYNTHETIC PATHWAY (Refer to pages 94 and 132)

The microsomal fractions from hamster intestinal mucosa and from rat liver are able to catalyse the formation of triglycerides from dihydroxyacetone phosphate (DHAP) and glyceraldehyde-3-phosphate (GAP) as well as from glycerol-3-phosphate (GP). The microsomes do not contain glycerol-3-phosphate dehydrogenase and therefore the formation of glycerides from DHAP or GAP need not proceed *via* their conversion into GP. Moreover, an inhibitor of triose phosphate isomerase completely inhibits glyceride synthesis from GAP but not from DHAP, indicating that DHAP is the immediate precursor of the glycerol moiety of the glyceride. The probable reaction sequence is:

$$\text{GAP} \rightleftharpoons \text{DHAP} \longrightarrow \text{acyl-DHAP} \xrightarrow{\text{NADPH}} \text{lysophosphatidate}$$

$$\xrightarrow{\text{acyl-CoA}} \text{phosphatidate} \longrightarrow \text{diglyceride} \longrightarrow \text{triglyceride}$$

(Rao, G.A., Sorrels, M.F., and Reiser, R. (1970) *Lipids* **5**, 762; Puleo, L.E., Rao, G.A. and Reiser, R. (1970) *Lipids* **5**, 770).

TRIGLYCERIDES: FORMATION OF ETHER BONDS

On pages 97 and 98, we described a scheme for the biosynthesis of ether bonds involving a condensation between a fatty alcohol and the carbonyl group of glyceraldehyde-3-phosphate (GAP). Recent work indicates that dihydroxyacetone phosphate (DHAP) and *not* GAP is the immediate precursor for the formation of the ether linkage since a specific inhibitor of triose phosphate isomerase inhibited the formation of an ether from GAP but not from DHAP. An ATP-dependent formation of a thiohemiacetal has been postulated to account for the ATP and CoASH requirements (Hajra, A.K. (1969) *Biochem. Biophys. Research Communs.* **37**, 486; Snyder, F., Malone, B., and Blank, M.L. (1970) *J. Biol. Chem.* **245**, 1790).

PHOSPHOLIPIDS: BIOSYNTHESIS OF PLASMALOGENS

It now seems certain that the vinyl ether group of plasmalogens (refer to pages 126, 127 and 134) is formed from the corresponding alkyl ether by biodehydrogenation. 1-Hexadecyl glycerol doubly labelled with tritium in the glycerol moiety and ^{14}C in the alkyl chain was converted by intact tumour cells into a plasmalogen which had the same $3H/^{14}C$ ratio as the precursor. One research group has isolated a compound which may be a 3-hydroxyhexadecyl glycerol suggesting that the mechanism of double bond formation involves hydroxylation and dehydration. However, the reaction has yet to be demonstrated *in vitro* and the mechanism remains speculative. (Wood, R., and Healy, K. (1970) *J. Biol. Chem.* **245**, 2640; Blank, M.L., Wykle, R.L., Piantadosi, C., and Snyder, F. (1970) *Biochim. Biophys. Acta.* **210**, 442).

GLYCOLIPIDS: TREATMENT OF LIPIDOSES BY ENZYME REPLACEMENT

For the first time, a lipid storage disease, Fabry's disease, (see page 162) has been treated by infusing patients with normal plasma. This provides active enzyme (ceramide trihexosidase) for hydrolysis of the substrate (galactosylglucosyl ceramide) which accumulates in the plasma of these patients. Maximum enzyme activity occurred 6 hours after infusion of the plasma and was detectable for 7 days. The accumulated substrate decreases about 50% on the tenth day after infusion (Mapes, C.A., Anderson, R.L., Sweeley, C.C., Desnick, R.J., and Krivit, W. (1970) *Science* **169**, 987).

General Index

ELUTION
of compounds from chromatograms, 7, 10, 12
EMDEN–MEYERHOF PATHWAY, 132
EMERGENCE TIME, in GLC, 13
EMULSION
formation of, in fat absorption, 104
formation of; physical chemistry, 177
of sterol esters and bile salts, 114, 115
of triglyceride as substrate for lipase, 98
ENERGY
activation, 78, 79
metabolism by mitochondria, 66, 73, 74
rich molecules, 65, 136, 139
storage as fats, 65, 104, 105
ENOYL–ACP
cis, trans isomers, 46, 47
reductase, 39, 46, 47
ENOYL–CoA
hydratase, 66–68
reductase, NADPH requiring, 43
ENZYME (See also individual enzymes)
activity, regulation of, 44, 45, 105
activity, dependence on lipids, 199
bile salt complex, 114
bound intermediate, 41, 49, 76
Commission, 4, 140
complex, 39–47, 97, A
defects, in sphingolipidoses, 160–163
induction, 45
inhibition, inhibitors, (see inhibition)
membrane–bound, 55, 144, 192, 198, 199
nomenclature, 4
of β–oxidation, 67–70
of fatty acid synthesis, 35
replacement, 163, A
substrate interaction, 143, 182
turnover, 45
EPIMERASE
in β–oxidation of unsaturated fatty acid, 71
EPOXY
acid, 28, 30
resin, 194, 197
ERGOT OIL, 45
ESTER
methyl (see METHYL ESTER)
of carbohydrates, 168, 169
of inositol, 169
physical properties, 176
ESTERASE, 98, 120, 121
ESTERIFIED STEROL GLYCOSIDES, 113, 170
ETHANOL
ether mixtures in lipoprotein extraction, 181

production of acetyl–CoA from, 43
water-mixtures in lipoprotein fractionation, 183
ETHANOLAMINE
as phospholipid component, 127, 129
formation of, 132, 133
in base exchange reaction, 136
release from phospholipids by hydrolysis, 147
ETHER
diethyl, activation of phospholipases, 143
– as chromatographic solvent, 8, 84,
– in lipid extraction, 113, 175, 181
linkage, biosynthesis of, 97, 98, A – in
glycerides, 90, 91 – in hydrogen
bonding, 176 – in phospholipids, 129
– physical properties, 176 – vinyl,
(alkenyl), 129, 134, A
EVERTED SAC
technique in fat absorption, 114
EXCHANGE REACTION
of bases in phospholipids, 136, 143
of lipids between lipoproteins, 185, 186
EXTRACTION
of lipids from tissues, 5, 6, 144, 175
FAT
absorption, 105–112, 114
butter, mutton, wool, 23, 90
conversion into sugar, 67
depot, 104, 110, 184
effect in diet, 93, 110, 117, 187–189
free diet, effect of, 55
mobilization, transport, 104, 105, 182, 184
natural, composition of, 90–93
saturated, effect in diet, 117–119
turnover, 131
FATTY ACID, 21–87
activation of, 31, 33
albumin complex, 23, 104, 115, 182
amide linkage of, with sphingosine, 125, 129, 150
analysis, separation of, 8, 14–18, 80–87
bibliography, 87
biosynthesis, 33–35 (see individual types
below) – by elongation mechanism, 44 –
by reversal of β–oxidation, 43 – contrast
with β–oxidation, 44 – control of, 39, 44
branched chain, activation of, 31 – *anteiso*,
23 – biosynthesis of, 36, 41–43, – GLC
of, 80, 81 – *iso*, 23 – in Refsum's
disease, 76 – occurrence of, 23, 90, 110,
121, 169, 170 – oxidation of, 70, 71
chemical hydrolysis of from lipids, 6, 147
composition of, in bacterial glycolipids,
170 – in cell membranes, 191–192 – in
gangliosides, 154 – in glycerides, 90–93,

PALM OIL, 2
PANTOTHENATE, 32, 38
PARAFFIN (see hydrocarbon)
PARTITION (COEFFICIENT), 7, 100
PENICILLIN, 203
PENTOSE PHOSPHATE PATHWAY, 95
PEPTIDE
  bonds, 179
  moiety of lipoproteins, 184, 185
  tryptic, of phospholipase A, 141, 142
PERIODATE—SCHIFF REAGENT
  in detection of vicinal hydroxyl groups,
  146
PERMEABILITY
  of membrane, effect of lipid composition,
  191
PERMEASE, 192
PEROXIDASE, 75
PEROXIDATION, 78
PEROXIDES, 78
PETROSELENIC ACID, 47
PHASE
  pairs, in chromatography, 6—8
  reversed, 7, 146
PHENYL
  alanine, 179, 192
  groups in mycosides, 169, 170
  groups, labelling with, 65
  substituted fatty acids, 31
PHOSPHATE
  enzymic release of, 93, 94, 134, 143
  estimation of, 145
  independent turnover of in phospholipids,
  131, 143, 146
  inorganic, ^{32}P, 131, 132, 136
PHOSPHATIDATE PHOSPHOHYDROLASE,
  93—95, 134, 137, 143, 144
PHOSPHATIDE
  (See 'Phospholipid')
PHOSPHATIDE ACYL HYDROLASE
  (See Phospholipase A)
PHOSPHATIDIC ACID
  biosynthesis of, 93—96, 131, 132, 133
  breakdown to diglyceride, 94, 95, 133, 134,
  143, 144
  membrane-bound, 144
  role in phospholipid synthesis, 131—136
  role in triglyceride synthesis, 93—96, A
  structure, occurrence, 127
PHOSPHATIDYL CHOLINE (3—sn—
  phosphatidyl choline; lecithin)
  biosynthesis of, 132—134, 136, 137
  content of membranes, 190—192
  electrical charge of, 176
  formation of micelles by, 177, 178
  hydrolysis of, by phospholipases, 141—143

modification of fatty acid composition, by
  transacylation, 137
molecular species of, 130, 146
nomenclature, 3, 4
requirement of, for enzyme activity, 199
role of, in fat absorption, 112
structure, occurrence, 127
synthetic, use in proof of phospholipase
  specificity, 141
transfer of fatty acids from, to cholesterol,
  115—117
PHOSPHATIDYL CHOLINE: CHOLESTEROL
  ACYL TRANSFERASE, 115, 116
PHOSPHATIDYL CHOLINE: CHOLINE
  PHOSPHOHYDROLASE, 143
PHOSPHATIDYL CHOLINE: PHOSPHATIDE
  HYDROLASE, 143
PHOSPHATIDYL ETHANOLAMINE (3—sn—
  phosphatidyl ethanolamine)
  as substrate in cyclopropane fatty acid
  biosynthesis, 64
  biosynthesis of, 133, 134, 136
  chromatography of, 147
  electrical charge in, 176
  exchange of base in, 136
  involvement of, in lipopolysaccharide
  biosynthesis, 200, 203
  methylation of, to phosphatidyl choline,
  136, 137
  N—acetyl derivative of, 127
  reactivation of glucose—6—phosphatase by,
  199
  structure, occurrence, 127
  trivial name of, 2, 3
PHOSPHATIDYL GLUCOSE, 128
PHOSPHATIDYL GLYCEROL (3—sn—
  phosphatidyl glycerol)
  amino acyl esters of, biosynthesis, 133,
  138, 139 — amino acyl esters of,
  occurrence, 128, 138 — amino acyl
  esters of, structure, 128
  biosynthesis, 133, 135, 136
  detection of, by vicinal hydroxyls, 146
  glucosaminyl, 128, 139
  importance of, in chloroplasts, 166
  interaction of, with basic proteins, 181
  structure, occurrence, 128
  trans—3—hexadecenoic acid in, 25, 29, 51
  191, 192
PHOSPHATIDYL INOSITOL (3—sn—
  Phosphatidyl inositol)
  activation of phospholipase B by, 142
  as precursor of phosphatidyl inositol
  mannosides, 139
  biosynthesis of, 133, 135
  structure, occurrence, 127

VAN DER WAALS FORCES, 176, 177, 179, 181
VINYL ETHER LINKAGE
biosynthesis of, A
cleavage of, 143
occurrence in plasmalogens, 126, 129
VISION
importance of Vitamin A in, 119
VITAMIN, 34, 55, 78
A, — bibliography, 124 — damage by peroxidation, 78 — importance in vision, 119 — structure, 119
A, esters, 119, 120 — absorption, 120 — analysis, 123 — hydrolysis, 120
E, as antioxidant, 78
F, 55
WASHING lipid solutions to remove water-soluble contaminants, 5, 144
WATER
effect on lipid—lipid interactions, 177, 178
formation in hydroxylations, 77
hydrogen bonding of, 176
importance of, in TLC solvents, 145

metabolism, essential fatty acids and, 55
permeability of skin, essential fatty acids and, 56
soluble substances, in lipid solution, 144
structure, fatty acids and, 30
tetrahedral structure of, 176
use of, as spray in TLC, 146
WAX
as component of surface lipids, 120, 121
complex bacterial (waxes A, B, C, D), 121, 169
definitions of, 120
ester, biosynthesis of, 121
hydrocarbon component, biosynthesis of, 121, 122
in capsular material of tubercle bacilli, 169
in cell wall of mycobacteria, 200
structure, occurrence, 120, 121
X-RAY CRYSTALLOGRAPHY, 58
X-RAY DIFFRACTION, 23, 177, 178, 194
ZONE, formation in chromatography, 7, 10, 17
ZWITTERIONIC, phospholipids, 177

SUPPLEMENTARY INDEX OF DISEASES AND DISORDERS OF METABOLISM

ATHEROSCLEROSIS, 117—119
ATAXIA, 162
CHRONIC POLYNEUROPATHY, 76
CORONARY ARTERY DISEASE, 117, 118, 187
DEGENERATIVE LESIONS, 117, 118
DIABETES MELLITUS, 45, 187
DISEASES OF LIPID METABOLISM (LIPIDOSES)
Lipoprotein metabolism, 187—189 — A—β—lipoproteinaemia, 188 — hypercholesterol-aemia, 188, 189 — hyperlipaemia, 110, 187 — hyperlipproteinaemia, 188 — hypo-lipoproteinaemia, 188 — Tangier disease, 188
Malabsorption syndromes, 110 — sprue, 110 steatorrhea, 110, 188
Refsum's disease, 76
Sphingolipidoses, 160—163 — cerebrosidosis, 162, 163 — Fabry, 162, 163 — Gaucher,

163 — metachromatic leucodystrophy, 162, 163 — neurovisceral gangliosidosis, 162, 163 — Niemann—Pick (sphingomyelinosis), 161, 163 — Tay—Sachs, 161—163
EGG WHITE INJURY, 34, 36
EPITHELIAL HYPERPLASIA, 56
FATTY LIVER, 187
HEPATOMEGALY, 162
ISCHAEMIC HEART DISEASE, 117—119, 189
KIDNEY DISEASE, 187
MYOCARDIAL INFARCTION, 117, 118
NEPHROSIS, 187
NIGHT BLINDNESS, 76
PINK—WHITE DISEASE, 51
PYREXIA, 162
RENAL FAILURE, 162
SKELETAL MALFORMATION, 76
SPLENOMEGALY, 162
XANTHOMATOSIS, 188

SUPPLEMENTARY INDEX OF SPECIES

SUPPLEMENTARY INDEX OF TISSUES AND SUBCELLAR PARTICLES